DIRT TO MILLIONS

MASTERING THE ART OF LAND DEVELOPMENT

GREG FARRICIELLI

LEGAL DISCLAIMER

This book is intended for **educational** and **informational purposes only**. While the author is a **licensed real estate professional**, the contents of this book **do not constitute legal, financial**, or **investment advice**. Readers should always consult with **attorneys, financial advisers, tax professionals, engineers, city planners, and other relevant experts** before making business or investment decisions.

Real estate development, land acquisition, and entitlement processes vary by location and are subject to **local**, **state**, and **federal laws** and **regulations**, which may change over time. **No guarantees are made regarding the accuracy, completeness**, or **applicability** of the information contained in this book to any specific project or market.

By reading this book, you acknowledge that all real estate transactions involve risk, and you agree that the author, publisher, and any affiliated parties **shall not be held liable for any direct** or **indirect losses, damages**, or **legal claims** resulting from the use of the information provided herein.

Your due diligence, experience, and professional judgment should always be the guiding factors in your real estate decisions.

The tools, resources, and platforms mentioned in this book are publicly available online and are included for informational purposes only. I do not endorse, guarantee, or receive compensation from any of these tools, nor do I make any representations regarding their accuracy, reliability, or effectiveness. It is always advisable to conduct your own research and due diligence before using any third-party services or software.

For permissions, inquiries, or rights-related matters, contact:

Greg@Stackedinvesting.com

Stackedinvesting.com

Published by:

Stacked Capital LLC

Nashville TN

ISBN: 979-8-9927075-0-2

CONTENTS

Introduction: Thank You for Taking This Journey
With Me!   5

1. WHAT IS LAND DEVELOPMENT EXACTLY?   15
Understanding Entitlement   15
Before You Start: Ask Yourself These Questions   16
The Value of Entitlement in Land Investing   17
Alternative Land Investment Strategies   17

2. LAND ACQUISITION STRATEGIES AND
MARKETING   23
Build the Right Infrastructure for Your Land Investment
Business   24
Build and Refine Your Cold Calling & Texting List   25
Direct Mail Marketing for Land Investing   31
Regulatory Compliance   34
Best Practices   34
Find Property Owners on Social Media   35
Paid Ads: A Broad Net with Mixed Results   37
The Power of Networking: Online and In-Person   39
Building a Buyers List   42
Managing Your Buyers List   45
Being Intentional About Your Networking   46
Building a Network of Contractors & Subcontractors   47
Finding & Nurturing Lender Relationships   48
Expanding Your Network Through Others   49
Final Thoughts: Your Network = Your Net Worth   50
Example Marketing Plan   51

3. MARKET RESEARCH & DUE DILIGENCE—THE
KEY TO PROFITABLE LAND DEALS   55
Common Exit Strategies for Land Development   57
Market Research: Analyzing the End Product and
Future Market Conditions   58
Determine Highest and Best Use   62
Using GIS & Municipal Data for Due Diligence   78

Understanding the Difference Between City and County
Jurisdictions   80
Understanding Municipal Codes and Regulations   82

4. LAND ENTITLEMENT & THE APPROVAL PROCESS   91
The Importance of Relationships with City Officials   92
Basic Entitlement Process   93
Legal & Regulatory Hurdles to Watch Out For   94
Team Members Needed for Success   96

5. CONTRACTS AND NEGOTIATION   105
Why We Saved Contract Negotiation for Last   105
Understanding Market Research: The Foundation of
Pricing and Demand   105
Due Diligence: Identifying Risks Before Making an
Offer   106
Entitlement and the Approval Process: Understanding
Timelines and Political Risks   108
Assembling the Right Team   109
Structuring the Right Exit Strategy   109
Bringing It All Together: Negotiation   110
Determining Contract Timelines Based on Approval
Processes and Timelines   111
Understanding of the Contract Parties   114
Understanding Key Contract Types for Land
Acquisitions   116
Navigating Contracts with the End Buyer   118
Working with Buyers Using Financing   121

Conclusion   125

CASE STUDIES: REAL DEALS, REAL LESSONS   127
Case Study: 503 W Trinity, Nashville, TN, 37207   127
Case Study 2: 411 W Trinity, Nashville TN, 37218
(Address Has Since Been Changed)   130
Case Study: The Three-Year Journey to Close 839 West
Trinity Nashville TN 37218 in Nashville—an Infill
Townhome Project   135

Epilogue: The Road to Mastery   141

References   145
About the Author   147

INTRODUCTION: THANK YOU FOR TAKING THIS JOURNEY WITH ME!

HOW I GOT INTO LAND DEVELOPMENT

At thirty-five years old, I was living in Nashville, Tennessee, working random restaurant jobs, playing guitar, going on small tours, and playing local gigs. I was making decent money, single, and living in my friend's basement for just $400 a month. Life was fun—a lot of fun —but I started feeling restless. The late nights were wearing on me, and I began questioning what my future looked like. Even though I had some savings and no debt, there was no real path to retirement.

Now, let me be clear—this isn't a religious book by any means, but I do believe that God speaks to us in different ways. And at that time, I felt led to get my real estate license. The funny thing is, I didn't even know any realtors. I had no clue what being a real estate agent really entailed. But something told me to take the leap.

Luckily, the job I had at the time allowed me 100 percent control of my schedule, so I enrolled in real estate classes. From the day I started studying to the day I had my license in hand, only about six weeks had passed. I was stoked! But then reality hit—I had no idea where to begin.

So, I turned to **Facebook**. I found a real estate investors' group in Nashville and started going to events. Early on, I knew that becoming an investor was the ultimate goal, but I needed to learn the business first. That led me to join a brokerage that I thought specialized in real estate investing.

Well, I quickly realized something was off. For one, I didn't like how the transactions were being handled. I kept noticing things that didn't add up, and I was expected to remember and repeat details that didn't seem true. It turned out that in between traditional real estate, they were "wholesaling."

For those unfamiliar, wholesaling is essentially getting a property under contract at a discount and assigning that contract to a third party for a profit. There's absolutely nothing wrong with wholesaling—I've done it hundreds of times—but the way I was being taught to do it wasn't right. On top of that, I later realized I was being paid as a "team member" and not as a real estate agent, meaning my commission split was much lower than I had thought.

Anyway, onward!

MY FIRST BIG DEAL

Around 2015–16, the real estate market was starting to recover from the 2009 crash. Prices were rising, and Nashville was booming. Builders were tearing down old houses and putting up two homes in their place, all because of existing zoning that didn't have value until this time.

At that time, I was working a mix of residential real estate deals and rental properties when a partner from Florida, who was working with my brokerage, sent me a lead. I did what I always did—I called the seller. I ended up getting three parcels under contract for **$20,000 each**.

Now, I had no idea what I was sitting on. But I started looking at the zoning and realized that those three parcels weren't just three lots—

they were **six buildable pads**. So, I threw together a quick investor write-up and just took a wild guess at the pricing based on the market. I listed the pads at **$30,000 each**, meaning each **lot** was now worth **$60,000**.

I didn't know what to expect. But by the end of the **same day**, I had multiple offers.

I ended up selling four pads to one investor and two to another, making **$180,000** from a **$60,000 contract**. We were blown away! To put this in perspective, those same six pads today would be worth between **$500,000** and **$600,000!**

One of the buyers for two pads backed out, but a real estate agent brought me another buyer, and everything closed in just a few weeks. The company I worked for made **$120,000** in profit, and I got my share. At the time, most people made that kind of money in a **year**, and I had made it in **one deal** with minimal effort!

That was the moment I realized: **I was onto something**.

After that deal, I left my brokerage and joined a more traditional real estate company. But the funny thing was—no one there really understood what I was doing. That worked to my advantage.

Fast forward, and I got a call from a guy who had a parcel of land that he had been working on for almost a year, trying to clear up title issues. He wanted **$175,000** for it, and he knew the zoning allowed **twelve units** to be built on it.

I gave him two options:

1. I could put the lot under contract and wholesale to an end buyer and get my fee that way, or …
2. I could list it on the retail market and see what happened.

He opted for the contract route, so I locked it up. A few days later, he called back and told me he actually had nine more months of title work left to do. I figured the deal was dead.

Nine months later—out of nowhere—he calls me again and says, "Are you still interested?" Of course I was! I put the contract together, and then I did something different this time. Instead of just marking the lot for the assignment fee (wholesale fee), I structured **two purchase options** that were a win-win either way for both parties:

- If the buyer agreed to let me list all twelve homes once built, I'd take **$5,000** upfront as a quick fee.
- If they wanted to buy me out of the listings, I'd take **$25,000** instead.

They chose to keep me as the listing agent, and when everything closed, I not only made money on the land deal—I also got **twelve sales** from it!

Then, we bought two more lots across the street, which turned into another couple of upfront assignment fees and thirteen more listings. Two of those homes even resold a few years later, bringing the total to twenty-eight transactions—all from that one deal!

Over time, I built relationships with the builders who bought the twelve units we just sold, and they started coming to me for advice. They wanted to know if they should sell the dirt or build and sell houses.

By this time, Nashville's land prices were appreciating so fast that sometimes it made more sense to sell the land without even building. I started listing **dirt** at prices that seemed ridiculous to me, but it **kept selling**.

From there, I became **not just a land broker but also an investor**. I started **funding deals** for them, earning interest or equity on my money while still getting the real estate listings.

For the next **four to five years**, we built and sold **around 100 homes per year**.

SHIFTING GEARS: THE BIRTH OF LAND DEVELOPMENT

By **2021**, after years of building homes at a massive scale, we were exhausted. Managing forty homes at a time meant juggling:

- 40 real estate agents
- 40 sellers
- 40 buyers
- 40 lenders
- 40 inspections
- 40 appraisals

It was a big PAIN IN THE ASS!

That's when we made the decision: **Let's just focus on entitlements**.

Instead of building houses, we started taking raw land, getting it **entitled and ready for development**, and selling it to builders.

Since 2021, we've **assembled and sold land from anywhere between 10 and 500 units at a time**, typically in the **60–80 unit range**, making **three to four times** our investment.

We knew, without a doubt—**this was the path forward**.

Before we get into the nuts and bolts, I want you to open your mind and know that anyone can do this, which is why most people don't, and how you can pivot into land development.

When people think of real estate investing, they typically think of flipping houses, buying rental properties, or wholesaling single-family homes. These are the most common and widely taught strategies, primarily because they offer quicker returns and are easier to understand. The idea of land development, on the other hand, can seem intimidating. But the truth is, anyone who understands real estate fundamentals can transition into land development—it's just a different set of skills and a longer timeline.

THE BIGGEST BARRIER: TIME, UNCERTAINTY, AND RISK

Most people shy away from land development for three reasons:

1. **It takes longer.** Unlike wholesaling or flipping a house, where you can see a return in a few months, land deals can take years to reach full profitability. The process involves zoning changes, government approvals, engineering, and infrastructure planning. Many people simply don't have the patience or long-term vision.
2. **Uncertainty is high.** With a house, you can see what you're buying and assess its value based on nearby sales. With land, you're dealing with potential—not existing value. There's no guarantee the city will approve your plans, that buyers will materialize, or that development costs won't exceed expectations.
3. **It requires capital and knowledge.** Land investing can require huge amounts of upfront money. You will need liquidity for contract deposits, engineering docs, and, depending on the situation, pay for certain studies (we will get to that later). It also requires patience and the ability to work through the red tape of city approvals, zoning codes, and legal constraints.

These three factors are why most real estate investors stick to traditional methods. But that's also why the opportunity in land is so much greater—because fewer people are competing in this space.

Time and time again, people reach out wanting to work with me on land deals, excited about the potential but hesitant about the uncertainty. The first question they always ask is, "How much will I get paid, and when?" And the honest answer is: **I don't know!** That's the reality of land development—there's no set paycheck, no guaranteed timeline, and no blueprint that works the same for every deal. Some projects close in months, and others take years. Some deals bring in six figures, while others break even or fall apart entirely.

The risk is what scares most people off. They don't realize that in land development, compensation isn't just about showing up—it's about what you're willing to risk. Putting up capital, signing on debt, taking on legal and financial responsibility, and doing the work to move a project forward—those are the factors that contribute to your payday. And the deal structure constantly evolves. The numbers you see in the beginning often look completely different by the time it closes. Entitlement costs change, lenders shift terms, partnerships adjust, and the city throws unexpected hurdles your way. You might start out thinking you're getting paid one way, and by the end of the deal, it's structured entirely differently.

This uncertainty is why most people don't make it in land development. They want a guaranteed return for minimal effort, but that's not how this business works. It rewards those who can adapt, take calculated risks, and see the bigger picture. And I can't count how many times those same people who walked away later regret it, watching from the sidelines as deals they could have been part of close for millions. The biggest hurdle in this business isn't finding opportunities —it's having the patience and vision to stick with them long enough to see the reward.

HOW TRADITIONAL REAL ESTATE INVESTORS AND AGENTS CAN PIVOT INTO LAND DEVELOPMENT

Most real estate agents are stuck in the rat race, fighting over the same listings and buyers, all for a measly 3 percent commission per deal. The amount of time, effort, and energy that goes into closing a single transaction—endless showings, client hand-holding, negotiations, and paperwork—often makes it feel like you're working for free until closing day. Now, let's put this into perspective: If you make just

$100,000 on a land development deal, that's the equivalent of an agent selling **$3,333,333** in real estate volume to earn the same commission! That's **ten to fifteen houses** in many markets! Instead of grinding out **forty to fifty deals a year**, what if you could close just **two or three high-margin deals** and make the same amount—without the constant hustle of competing for listings and dealing with emotional buyers? Land development allows you to control your deals, work at a higher level, and **build real wealth** instead of just chasing commissions.

If you've ever done wholesaling, fix-and-flip, rentals, or commercial deals, you already have the foundation needed to move into land development. The transition is easier than you think.

1. **If You're a Wholesaler:** Instead of flipping contracts on houses, you can start flipping land contracts. Many of the same marketing and acquisition strategies apply—direct mail, cold calling, and working with distressed sellers—but instead of looking for rundown houses, you're looking for underutilized land with development potential.

2. **If You're a Fix-and-Flip Investor:** Instead of renovating houses, you're repositioning land. You take an undervalued piece of dirt, entitle it for higher density or a more valuable use, and sell it to a builder just like a house flipper sells a rehabbed home.

3. **If You're a Buy-and-Hold Investor:** Instead of cash flow from rental properties, you can earn passive income through land banking, owner-financing land deals, or participating in long-term appreciation.

4. **If You're a Commercial Investor:** You already understand how zoning, location, and infrastructure impact value. Moving from commercial real estate to mixed-use or land development is a natural extension.

WHY THE LONG TIMELINE IS ACTUALLY AN ADVANTAGE

While most people view the long development process as a negative, smart investors see it as an advantage.

1. **Less Competition:** Because fewer investors have the patience to wait for entitlements and zoning changes, there's less competition in land deals than in house flipping or multifamily investing.
2. **Massive Upside Potential:** While house flippers might make a 20 percent return on a rehab, land developers often triple or quadruple their investment because they are adding real value by securing entitlements, engineering approvals, and rezoning.
3. **Deals Don't Require Immediate Capital:** In many cases, land deals are structured with long due diligence periods and contingencies, allowing you to secure control of the land before needing major upfront capital. This allows smaller investors to play in a space that seems reserved for big developers.

THE MINDSET SHIFT: THINKING LIKE A LAND DEVELOPER

The biggest change required to succeed in land development isn't money or experience—it's mindset. You have to move from thinking about quick profits to long-term value creation. You're not just buying a piece of dirt—you're unlocking its highest and best use through entitlements, zoning, and planning.

Yes, land development takes longer. Yes, it comes with more uncertainty. But that's exactly why it's one of the most lucrative and underutilized strategies in real estate. The best part? Most of your competition won't even try.

If you're willing to learn, adapt, and play the long game, you can build wealth in land development just like you would in any other area of real estate. And with the right knowledge and strategies, you can start right now.

Let's jump in!

CHAPTER 1
WHAT IS LAND DEVELOPMENT EXACTLY?

When most people hear the term "developer," they envision large construction companies erecting skyscrapers or sprawling subdivisions. While this image isn't incorrect, it represents just one facet of development. In this book, when I refer to development, I'm focusing on a crucial phase that precedes construction: transforming raw land into "build-ready" or entitled land.

UNDERSTANDING ENTITLEMENT

Land development is a process that involves the evaluation, planning, engineering, codes, and construction to bring the highest and best value to the marketplace after improvements, based on codes and regulations set by the municipality and regulatory agencies. The entitlement process determines:

1. **Permissible Uses:** What can be built—residential homes, commercial buildings, mixed-use developments, etc.
2. **Density and Size:** How many units or structures can occupy the land.
3. **Infrastructure Requirements:** Necessary provisions like

roads, sewage systems, water lines, and environmental considerations.

4. **Regulatory Constraints:** Zoning laws, environmental regulations, and other legal restrictions.

For instance, owning a ten-acre plot doesn't automatically grant permission to build a hundred homes. Developers must navigate zoning laws, environmental assessments, and community hearings to secure the right to develop the land as envisioned.

BEFORE YOU START: ASK YOURSELF THESE QUESTIONS

Land investing isn't like flipping houses. It takes patience, persistence, and problem-solving. Before you get started, ask yourself these three crucial questions:

- **Question #1:** Are you ready for a delayed but profitable payday? Can you handle the grueling process of working with city officials and regulatory agencies? Do you have funds set aside for earnest money and due diligence?
- **Question #2:** Are you prepared to have educated conversations with sellers, justify your offers, and negotiate long-term deals with follow-ups that could take months (or even years)?
- **Question #3:** Can you hold yourself accountable for giving buyers accurate, high-quality information about the land you're selling? One bad assumption or misleading statement could ruin your reputation and cost you a fortune.

If you answered YES to these, then land investing may be for you. If not, you may want to consider another niche in real estate before jumping in.

THE VALUE OF ENTITLEMENT IN LAND INVESTING

Entitling land significantly enhances its value. Raw land might appraise at $300,000, but once entitled for a forty-unit residential project, its worth could soar to $2 million or more depending on demand and location. This increase stems from the reduced risk and effort for builders, who prefer purchasing land primed for construction over undertaking the lengthy and uncertain entitlement process themselves.

As land developers, our role is to identify promising raw land, navigate the entitlement process to unlock its potential, and then sell it to builders at a premium. This approach doesn't involve physical construction but focuses on adding value through strategic planning and approvals.

ALTERNATIVE LAND INVESTMENT STRATEGIES

Beyond entitlements, several other lucrative land investment strategies exist. It's good to be familiar with these so you can recognize any opportunities.

1. **Land Banking:** Land banking entails purchasing land in the path of anticipated growth and holding on to it until its value appreciates due to development pressures. A notable example is Walt Disney's acquisition of over 27,000 acres in Florida during the 1960s. By secretly purchasing vast tracts through various entities, Disney secured the land at low prices before announcing plans for Disney World, which led to exponential land value appreciation. Investors today use this strategy by acquiring land on the outskirts of growing cities, expecting urban expansion to drive up its worth over time.
2. **Land Flipping:** Land flipping involves buying land at a lower price and selling it at a higher price within a short timeframe, often after making minor improvements or securing favorable

entitlements. For instance, an investor might purchase a neglected parcel, clear debris, and obtain zoning changes to enhance its marketability before reselling at a profit. This strategy allows for quick turnarounds and profits without the long holding periods associated with land banking.

3. **Land Flipping with Owner Financing:** A variation of traditional land flipping, this strategy involves purchasing undervalued land and selling it at a higher price while offering financing to buyers who may not qualify for traditional loans. By providing owner financing, investors expand the pool of potential buyers and create passive income through interest on the financed amount. For example, an investor might acquire a rural plot for $10,000 and resell it for $20,000 with a $1,000 down payment and monthly installment plans, making the land more accessible while increasing profitability.

4. **Subdividing Land:** This strategy involves purchasing a large parcel and dividing it into smaller lots for resale. Since smaller tracts often command higher prices per acre, subdivision can significantly increase the total return on investment. For instance, an investor might acquire a ten-acre property, subdivide it into ten one-acre lots, and sell each lot individually to home builders or individual buyers. Successful subdivision requires thorough research into local zoning laws, infrastructure requirements, and market demand.

5. **Covered Land Plays:** A covered land play involves acquiring income-producing properties with the intention of redeveloping them in the future. For example, an investor might purchase a parking lot in an urban area that generates modest income while waiting for zoning changes or market conditions to allow for a more profitable redevelopment, such as a mixed-use building. This strategy balances immediate income with long-term redevelopment potential, ensuring steady cash flow while positioning the land for higher future value.

Each of these strategies offers unique opportunities and challenges. The key is to assess your resources, risk tolerance, and market knowledge to determine which approach aligns best with your investment goals. Now, let's get into larger development plays, which are the focus of this book.

In the realm of land development, various entitlement projects cater to different market needs and urban planning goals. Below are detailed definitions of key project types, examples from Nashville, and insights into the markets where they thrive:

1. *Residential Infill*

Definition: Residential infill involves developing vacant or underutilized parcels within existing urban areas, aiming to enhance density and make efficient use of existing infrastructure.

Example: In Nashville, the Eastside Infill project exemplifies this approach, where modern homes are constructed in established neighborhoods, seamlessly integrating with the existing community fabric.

Optimal Markets: This strategy is most effective in urban centers experiencing population growth and housing demand, where available land is scarce and there's a premium on proximity to amenities and workplaces.

2. *Residential Single-Family Subdivisions*

Definition: This entails developing a tract of land into multiple single-family homes, often featuring shared amenities like parks and community centers.

Example: The Magnolia Grove Reserve in Murfreesboro, near Nashville, showcases a planned community offering a range of home designs catering to families seeking suburban living with convenient access to urban centers.

Optimal Markets: Ideal for suburban areas adjacent to growing cities, appealing to families desiring more space, affordability, and access to quality schools.

3. *Multifamily Projects*

Definition: These developments consist of multiple residential units within a single building or complex, accommodating several families separately.

Subtypes:

- **Wrap Buildings:** A large multifamily structure encircling a multi-level parking garage, optimizing land use and providing residents with convenient parking access.
- *Example:* Nashville's River House is a notable wrap building offering luxury apartments with integrated parking solutions.
- **Garden Apartments:** Low-rise complexes, typically two to three stories, surrounded by landscaped grounds, offering a blend of affordability and community feel.
- *Example:* Haynes Garden Apartments in Nashville provides residents with comfortable living spaces amid green surroundings.

Optimal Markets: These projects are suited for urban and suburban areas with high rental demand, particularly where there's a need for diverse housing options to accommodate varying income levels and lifestyle preferences.

4. *Mixed-Use Developments*

Definition: Projects that combine residential, commercial, and sometimes industrial spaces within a single development, fostering a live-work-play environment.

Example: The River North development in Nashville is a prime example, integrating housing, retail, and office spaces to create a vibrant community hub.

Optimal Markets: Thrives in urban settings with a push toward sustainable living, appealing to residents and businesses seeking convenience and a dynamic environment.

5. *Commercial Projects*

Definition: Developments focused on business activities, including office buildings, retail centers, hotels, and restaurants.

Example: The Fifth + Broadway complex in downtown Nashville encompasses retail spaces, dining establishments, and office suites, contributing significantly to the city's economic landscape.

Optimal Markets: Best suited for bustling urban centers or rapidly growing suburban areas with strong economic activity, catering to businesses and consumers alike. Understanding these project types and their ideal market conditions is crucial for developers aiming to align their investments with community needs and market demands.

CHAPTER 2
LAND ACQUISITION STRATEGIES AND MARKETING

I n the world of land development, the best deals aren't always on the MLS (but they can be). Some land deals can be found on the MLS because many real estate agents who list land don't fully understand how to price or market it properly. This lack of knowledge can lead to undervalued listings, misclassified zoning descriptions, or missed entitlement opportunities, creating an advantage for investors who know what to look for—they're secured long before they ever hit the open market. The key to success in this business is knowing where to look, how to structure deals, and how to position yourself in front of sellers before anyone else.

Most investors get stuck because they don't know where to start. They assume that land is bought and sold like traditional real estate. It's not. The best land deals are often hidden, sitting in the hands of owners who don't even realize their land has value—or worse, they do know but have no idea how to navigate the entitlement process.

This chapter will cover exactly how to find and secure off-market land deals before your competition, including direct-to-owner marketing, data tools, negotiation strategies, and creative financing, but let's get set up first.

BUILD THE RIGHT INFRASTRUCTURE FOR YOUR LAND INVESTMENT BUSINESS

Before launching any marketing efforts, it's **critical to establish a strong infrastructure** that presents your business as credible and professional. Many landowners, builders, and investors will **look you up online before responding to your calls, texts, or mailers**. If you don't have a visible presence, you risk losing potential deals before the conversation even starts.

Key Components of a Solid Business Infrastructure:

1. **Professional Website:** Your website serves as your **digital storefront**. It should clearly state:
   - Who you are and what you do
   - The niche you specialize in (e.g., **land acquisition, entitlement, infill development**, etc.)
   - A way for property owners to submit their land information for a potential offer
   - Testimonials, past deals, or credibility markers
2. **Branded Social Media Pages:** Create business pages on **Facebook, LinkedIn**, and **Instagram** with:
   - Your logo and branding
   - Consistent posts about your land deals, market insights, or behind-the-scenes content
   - Contact information that matches your website
3. **Main Business Phone Number:** Have **one dedicated number** (not your personal cell) that is used for **all marketing efforts**. Use a service like **CallRail, Google Voice**, or **OpenPhone** to track calls and texts efficiently. This is separate from your marketing phone numbers that you use for mail and calls.
4. **Clear Business Positioning:** Make sure your branding **clearly communicates your niche**. Whether you focus on **entitled**

land, infill lots, farmland, commercial development, or **mobile home lots**, your messaging should make it obvious.

5. **CRM (Customer Relationship Management) & Lead Tracking System:** All incoming leads should be **organized in a CRM** (like **REsimpli, Podio, HubSpot**, etc.) to track conversations, follow-ups, and deal progress.

Why This Matters

Without the right infrastructure, your marketing **lacks credibility**. Having a **professional online presence** and a **structured system for handling leads** will put you miles ahead of competitors who rely on outdated, unorganized methods. So, **before you drop a dime on direct mail, cold calling, or paid ads, make sure your business is set up to capture and convert leads like a pro**. The last thing you want is to have a great conversation or get a seller to respond—only for them to look you up and find ... nothing. To them, you're just another spammy scammer, and let's be real—you're not spam. You're a rockstar who does business with integrity! Invest the time and resources upfront to build your credibility, and you'll make everything run smoother when those leads start rolling in!

Now that you have infrastructure in place, it's time to:

BUILD AND REFINE YOUR COLD CALLING & TEXTING LIST

Before making a single call or sending a text, the first step is **pulling a high-quality list and skip tracing it for accurate contact information**. The success of your outreach campaigns depends largely on the accuracy and depth of your data.

Unlike traditional real estate, where you can browse the MLS and Zillow for listings, **the best land deals require proactive deal-finding strategies**. This means you're often going straight to landowners before they've even thought about selling.

The **first step in any successful cold calling or texting campaign is pulling a high-quality list of landowners**. Many people jump straight into outreach without taking the time to ensure their data is accurate, which leads to wasted effort and money. Using robust data sources like **DataTree, PropStream**, and **LandVision** allows you to filter and refine your list based on specific criteria relevant to your investment strategy.

DataTree, for example, offers a vast amount of property information, including owner details, transaction history, zoning classifications, assessed values, and even mortgage data. The ability to **narrow down your list** based on **property type, acreage, zoning, tax status**, and **ownership duration** makes your marketing more targeted and efficient.

Before launching a campaign, it's critical to **scrub your list, remove duplicate records**, and **filter out properties that don't fit your criteria**. You can also cross-reference your list with **GIS (geographic information system) parcel maps, county tax records**, and **local planning department data** to ensure accuracy. The more refined your list is upfront, the better your response rates will be when you begin making calls, sending texts, or launching direct mail campaigns. Be sure to pull out any land that may belong to schools, roads, easements, or anything that you don't want to target. Let's get into the marketing options:

Where to Pull Lists

I don't endorse any of these companies so make sure you do your own due diligence, but these are some of the sources that you can find online.

- **PropStream:** Provides detailed property data, including owner information, mortgage status, and transaction history.
- **LandVision:** Great for large-scale land acquisitions and development projects.

- **Reonomy:** Ideal for commercial land and multifamily asset searches.
- **County Tax Assessor & GIS Websites:** Public records can provide information on tax delinquencies, estate sales, and ownership history.

Once you have a list, **skip tracing** is the next step. This is the process of obtaining **phone numbers**, **email addresses**, and sometimes even **social media profiles** for the property owners.

Keep in mind:

Land often appears in records with **no assigned address** or a **"0" address** because it hasn't been developed or officially recorded with a physical location. Unlike houses or commercial buildings, raw land may only be identified by **parcel numbers (APN)**, **lot numbers**, or **legal descriptions**. This is common in rural areas, new developments, or landlocked properties. When searching for these properties, it's important to use **GIS mapping tools**, **tax assessor records**, or **neighboring addresses** to pinpoint their exact location. Always ask the owner for a **parcel ID**, **APN**, **neighboring address**, **name on tax record**, or **the nearest cross street** to ensure you're looking at the correct property.

Skip Tracing & Identifying the Right Contacts

Not all skip tracing services are created equal. Some **only pull publicly available phone numbers**, while others **can dig deeper to reveal LLC owners, family members, or business partners**.

Some skip tracing services include:

- **Batch Skip Tracing:** Can uncover LLC owners and their personal contact details.
- **Direct Skip:** Provides additional data like associated relatives and past addresses.

- **TLOxp (TransUnion):** High-end service often used for deep-dive research.
- **DataTree.** Provides nationwide property data.

Refining Your List

The first round of calling or texting is primarily about filtering out bad numbers. You should:

1. **Flag disconnected numbers** and remove them from your list.
2. **Identify good numbers** and mark them as high-priority for follow-up.
3. **Find alternate contacts** for hard-to-reach owners, including **family members**, **partners**, or **business associates** (some skip tracing services provide this).
4. **Use social media** to track down owners when necessary.

Don't get discouraged if the first round doesn't yield immediate results—the real goal is to **clean your list, refine it, and create a CRM system with verified contacts.** Some owners might not answer on the first attempt but will later recognize your number and respond after repeated contact.

Making Calls & Sending Texts Effectively

- **Use a dedicated number that doesn't show up as spam.** Services like CallRail allow you to set up multiple local numbers to increase response rates.
- **Always use a local area code for texting and calling.** This builds trust and makes it more likely that people will answer.
- **Change numbers after running through a list to maintain deliverability and avoid being flagged.**
- **Track every response in your CRM,** noting which numbers work and which owners prefer texts over calls.

- **Don't use a Virtual Assistant (VA) for cold calling in land acquisition**—it requires **a more nuanced conversation** and experience handling objections.

By following this structured approach, **you're not just reaching out blindly—you're strategically building a high-quality list of verified owners and setting yourself up for future follow-ups.**

Customizing Your Cold Call Scripts for Maximum Effectiveness

Having a **well-structured script** is crucial for cold calling, but one-size-fits-all doesn't work when it comes to land investing. **Different markets require different questions**, so it's important to tailor your script to gather the most valuable information for each area. Be sure to use the example I provided and tweak it to your market and area.

For example, in **Florida**, a key factor for buyers is **waterfront access**. If a property is near the water, you'll need to ask whether it has a **seawall, a dock,** or **direct ocean access**—details that significantly impact the value. In contrast, when calling about **infill lots in Tennessee**, you'll want to confirm **alley access, existing utilities**, and **potential zoning restrictions** that could affect buildability.

Another crucial aspect of a strong script is how you initiate the conversation. I always assume familiarity with the seller, so I begin the call by saying their first name with an upward tone—this creates the impression that I already know them. If it turns out to be the wrong person, I promptly ask if they know the actual owner, avoiding unnecessary delays.. **Never start with generic pleasantries like "How are you?" or "How's it going?"**—this gives the seller an easy out to hang up. Instead, get straight to the point.

Your script should also include a section to **dig deep into property details** beyond what's publicly available, uncover the **seller's motivation**, and confirm **ownership status**. Asking the right questions early can save you from wasting time on properties with hidden issues.

If the seller mentions they have a **real estate agent**, don't immediately shut down the conversation. Instead, **ask for the agent's name and contact information** while still gathering as much property information as possible from the seller. Some sellers will still give details, even if they have an agent representing them. If they insist that all communication go through their agent, respect that and **reach out to the agent directly**—but be sure to mention that you already spoke with the seller to establish credibility. Sometimes, agents may have additional insights about the seller's motivation or flexibility on price, so maintaining a professional and cooperative approach can keep the deal moving forward.

When texting potential sellers, **avoid addressing them by name upfront** because skip tracing data isn't always accurate, and you may have the wrong contact. Instead, structure your message around the property itself, like: **"Hi, are you the owner of [property address]?"** This approach feels more natural and avoids immediate resistance if the recipient isn't the actual owner. If they respond, you can then confirm ownership and continue the conversation. Once you know whether they are the owner or not, you can ask if they've had thoughts of selling or developing or whatever you want. I don't like to ask questions until I know who I'm talking to.

Finally, **handling objections and friction points smoothly** is just as important as the main script itself. Have quick, confident answers ready for common pushbacks, such as **"How did you get my number?"** or **"I have a realtor."** Having a **clear and direct script**, along with a well-thought-out follow-up system, ensures that you're maximizing your time and increasing the likelihood of closing a deal.

If you want a copy of the script, If you want my calling scripts send me an email!

DIRECT MAIL MARKETING FOR LAND INVESTING

Direct mail is one of the most effective ways to generate leads in land investing, but it also comes with one of the highest marketing costs. Investors should expect to spend anywhere between **$500 to $5,000 per campaign**, with **response rates typically ranging from 0.5 percent to 3 percent**.

This means that out of 1,000 mailers, you may only get 5–30 responses, making consistency and follow-ups crucial. The best approach is to **mail weekly or biweekly for the first two months**, then scale back to **once per month for continued engagement**.

There are three primary types of direct mail strategies that land investors use:

1. **Blind Offers** or **Letter of Intent:** These include an **actual purchase offer** in the letter. This strategy can work well for highly targeted lists where the investor has a solid understanding of the market. These are used mostly for land flipping rather than entitlements, since they're hard to price and the seller may not understand your valuation. Even if it's a large number, they could ignore it or not take you seriously. These are better for rural land where there is either a slow-moving market or some type of distress.

2. **Neutral Letters:** These are **softer** letters simply stating interest in purchasing the land, encouraging the seller to reach out without an explicit offer. These are great for gaining credibility and introducing your company. For land entitlements, these seem to work best but are not guaranteed.

3. **Postcard:** A compact, visually appealing mail piece used in real estate marketing to grab attention quickly. It's particularly effective for **brand recognition, generating inbound leads**, and **staying top-of-mind with sellers**.

Each strategy has its advantages: **Blind offers can help weed out unmotivated sellers**, while **neutral letters generate more responses and allow for negotiation flexibility**, and post cards are for **brand recognition**. I suggest trying all.

Best Practices for Mailing

When conducting a direct mail campaign, one of the most critical steps is scrubbing the list before sending out thousands of mailers. A VA or team member should go through each property and apply specific filters that match the target market area. Important parameters can include but are not limited to:

1. **Sewer vs. Septic:** Sewer access is often more valuable and buildable.
2. **Landlocked Properties:** Avoid properties with no access to roads unless you plan to secure easements.
3. **Zoning Restrictions:** Ensure the land aligns with your development goals.
4. **Flood Zones and Topography Issues:** Avoid major environmental restrictions.
5. **Lot Size and Subdivision Potential:** Ensure parcels meet local zoning requirements.

This upfront work ensures you're not wasting time and money marketing to properties that don't fit your criteria. Once the list is refined, it's beneficial to organize it into a **spreadsheet** with relevant details so that when sellers call, you can quickly assess the property and respond intelligently. You can usually find a VA that will scrub for around ten to twenty cents per record.

Having a **dedicated phone number** and **landing page** for the mail campaign allows sellers to learn more about you and ensures a stream-lined intake process. Many investors use an **answering service like PAT Live**, which provides a professional touch while filtering out low-quality leads. A well-designed **script** for the answering service should be **under a minute**, covering essential details like the caller's name, property location, and motivation for selling. This allows you to call back only after gathering enough preliminary data, ensuring more productive conversations.

Keeping track of all inbound calls in a **separate spreadsheet** helps refine future campaigns. By tracking what types of properties generate the best responses, you can fine-tune your mailing criteria, adjust your offer pricing, and maximize conversion rates. **Data is key**, and every campaign should be analyzed for patterns and improvements.

To maintain professionalism and streamline operations, **a PO box or virtual mailbox like Anytime Mailbox** is highly recommended. A PO box provides privacy, while a virtual mailbox allows mail to be **scanned and emailed to you as PDFs**, eliminating the need to physically check a mailbox. This can be especially helpful when handling **returned mail**, processing seller inquiries, and organizing your campaign data.

Text & Voicemail Follow-Ups for Mail

Direct mail should be complemented with **follow-up texts and calls** to reinforce branding and credibility. Many sellers ignore mail until they receive a text or call. However, it's best **not to use VAs for outbound calls** in land acquisition, as this is a competitive field where you're often dealing with experienced investors and high-value sellers. Unlike house wholesaling, where automation works well, land sellers expect **experienced professionals who can talk numbers immediately**.

Here is an example script. Remember to tweak this to fit your business and see how I've added every lead source scenario like whether they've responded to text, call, or mailer.

When implementing cold calling and texting strategies in real estate investing, it's crucial to adhere to legal regulations and employ best practices to maximize effectiveness and minimize risks. I have a friend who had a $20,000 fine for a text! Be careful. There are mean people out there who just want to take advantage of hard-working investors!

REGULATORY COMPLIANCE

The **Telephone Consumer Protection Act (TCPA)** governs telemarketing activities, including cold calling and texting. To remain compliant:

- **Obtain Consent:** Before sending promotional text messages, secure express written consent from recipients.
- **Respect the National Do Not Call Registry:** Scrub your call lists against the registry to avoid contacting individuals who have opted out of telemarketing communications.
- **Provide Opt-Out Mechanisms:** Include clear instructions in your messages, such as "Reply STOP to unsubscribe," allowing recipients to easily opt out.

BEST PRACTICES

- **Dedicated Phone Numbers:** Utilize specific phone numbers for cold calling and texting to manage responses effectively. Ensure these numbers are registered appropriately to prevent being flagged as spam.
- **Local Area Codes:** Employ phone numbers with area codes matching your target market to increase answer rates and build trust with potential leads.
- **Personal Engagement:** Given the typically smaller size of land entitlement lists, consider making calls personally. This direct approach fosters trust and allows for nuanced discussions about property specifics.
- **Cautious Use of VAs:** While VAs can handle various tasks, cold calling requires in-depth knowledge and the ability to

address complex questions. Personal involvement is often more effective in these scenarios.

- **Separate Numbers for Texting:** Use distinct phone numbers for texting campaigns. After completing a round of outreach, consider changing numbers to maintain deliverability and avoid spam filters.
- **Customer Relationship Management (CRM) Systems:** Implement a CRM to track interactions, schedule follow-ups, and maintain organized records. Be aware that some CRMs have strict regulations regarding unsolicited communications; ensure your practices align with their policies.
- **Integration with Direct Mail:** Cold calling and texting can effectively complement direct mail campaigns. Following up on mailers with a call or text reinforces your message and can prompt recipients to engage with your offer.
- **Data Management:** Maintain detailed records of all communications. This practice not only aids in compliance but also provides valuable insights for refining future campaigns.

By adhering to these guidelines and staying informed about current regulations, you can execute cold calling and texting strategies that are both effective and compliant, enhancing your real estate investment efforts. Remember, this is a numbers game, and always know you have the right person before you give them any info (like your real name 😉).

FIND PROPERTY OWNERS ON SOCIAL MEDIA

Many landowners, especially in high-growth areas, are builders, investors, or developers who are **active on social media and profes-sional networking platforms**. When traditional contact methods like cold calling or direct mail don't work, social media can be an effective tool for **tracking down and engaging property owners directly**. Most of the companies will have a phone number or email on their page you can reach out to. Don't forget to google the companies and find their websites.

Where to Search for Owners on Social Media

1. **Facebook:** Many property owners, particularly smaller investors and builders, have personal profiles or business pages. Try searching their name, company, or even the parcel address in Facebook's search bar.
2. **Instagram:** Builders and investors often showcase projects here. Searching for their name or company can lead to direct contact.
3. **LinkedIn:** Larger developers, real estate professionals, and landowners frequently maintain LinkedIn profiles. This is a great place to connect with decision-makers professionally.
4. **Twitter/X:** Some investors and real estate professionals post deals and market updates here. If you find their account, you can engage with their content before reaching out.
5. **Real Estate and Land Forums:** Platforms like **BiggerPockets, Land Investors Forum**, and even **local investor Facebook groups** can be goldmines for finding and contacting owners.
6. **Skip Tracing Tools with Email Results:** Many skip tracing services (like **BatchSkipTracing, PropStream**, or **DataTree**) provide **emails** that can be used for outreach on social media.

Ways to Contact Owners on Social Media:

- **Direct Messaging (DMs):** If their profile allows messages, send a short and professional introduction explaining who you are and why you're reaching out. Keep it concise and to the point. You can also go to the websites of these prospects, find their social media links, and message them that way. That removes the gate keeper and takes you direct to the source.

- **Engagement Before Outreach:** If they post frequently, **like and comment on their content** before reaching out. This helps establish rapport.
- **Connection Requests:** On LinkedIn, sending a **personalized connection request** mentioning mutual interests or markets can increase the likelihood of a response.
- **Join Mutual Groups:** Many investors and builders are part of **real estate Facebook groups** or LinkedIn communities. Engage in discussions and **build credibility before pitching**.
- **Use Multiple Touchpoints:** If they don't respond on one platform, try another—**LinkedIn first, then email, then a direct message on Instagram or Facebook**.

If you don't see any Facebook groups specifically for connecting sellers, investors, and builders in your target market, create one yourself! A well-run networking group can position you as a local expert, attract motivated sellers, and bring in investors looking for opportunities. Start by building a group for industry professionals—developers, builders, and investors—where you can share insights, market trends, and off-market deals. At the same time, consider creating a neighborhood-specific group in your target area where local homeowners, landowners, and businesses can engage. By providing value, answering questions, and sharing useful information, you build trust and credibility. Over time, sellers will come to you first when they're ready to offload their land, and buyers will see you as a reliable source for development-ready properties. Plus, these groups can serve as a powerful marketing tool without the high cost of paid ads.

PAID ADS: A BROAD NET WITH MIXED RESULTS

Running **paid ads** (such as **Google Ads**, **Facebook Ads**, or **PPC campaigns**) can generate inbound leads, but in land investing, it often presents **significant challenges**. Unlike direct mail, where you target specific properties, paid ads **cast a wide net**, and filtering for vacant land **is extremely difficult**. This means that:

- You will receive leads for **all types of properties**—houses, commercial buildings, and land **that doesn't fit your criteria**.
- You may **get lucky** and land a good deal occasionally, but the overall return on investment **can be unpredictable**.
- The cost per lead **varies significantly** based on the market, keywords, and competition, but **expect to spend hundreds or even thousands per month** with uncertain results.

When Paid Ads Make Sense

If you're considering paid ads, it's best to **have a strategy in place** for handling the **leads that don't fit your land business**. Here's how you can **maximize your marketing spend**:

1. **Partner with an Agent or Investor:** Since many of the leads won't be land, you can **refer residential or commercial properties** to a trusted real estate agent or wholesaler. In exchange, you could negotiate:
   - A **referral fee** or marketing fee per closed deal.
   - A **percentage of the commission** if the agent closes the deal.
   - An arrangement where they **fund part of your ad spend** if they benefit from the leads.
2. **Only Use Paid Ads When You Have a Large Budget:** In the beginning, when your budget is **limited, direct mail, cold calling,** and **networking** offer **better return on investment (ROI)**. Paid ads should be used **later**, once your **lead funnel is stable and you can afford a long-term marketing strategy**.
3. **Target Tips to Improve Lead Quality:** If you decide to run paid ads, **fine-tune your targeting** to increase the chances of getting land leads:
   - Use keywords like **"sell my land fast"** instead of general real estate phrases.
   - Run ads in **specific counties or regions** where you invest.

○ Consider using **retargeting ads** to follow up with website visitors who previously showed interest in selling land.

Paid ads **can work**, but they are **not the most efficient** way to generate land deals. They **require a large budget, don't easily filter vacant land**, and often **produce low-quality leads**. If you **have extra marketing dollars** and a **referral system in place for unwanted leads**, then **paid ads can be a supplement** to your marketing—**but they should not be your primary lead source** in the early stages.

THE POWER OF NETWORKING: ONLINE AND IN-PERSON

One of the most **underrated yet powerful** ways to find land deals and build your reputation is through **networking**—both **online** and **in-person**. In this business, **your next deal could come from a relationship you build today**, whether it's with a **seller, investor, builder, broker, wholesaler**, or **city official**.

Online Networking: Tapping into Targeted Groups

In today's world, many of the best connections **start online**. Joining **real estate and land investing groups** on platforms like **Facebook, LinkedIn, and BiggerPockets** allows you to:

- **Connect with sellers** who may want to offload land but don't know where to start.
- **Meet buyers** who are actively looking for land and can be a future exit strategy for your deals.
- **Learn from other investors** about market trends, zoning changes, and off-market opportunities.
- **Find potential partners** for joint venture (JV) deals, wholesaling, or co-investing.

Engaging in **local groups** in your **target markets** is crucial. For example, if you are working in **Tennessee**, you should be in groups dedicated to **Nashville real estate, Chattanooga land development**, and **other active investor circles in the area**.

In-Person Networking: The Importance of Being on the Ground

While **online networking is a great first step**, nothing replaces **physically meeting people in your target market.** Attending **real estate meetups, builder associations**, and **land development conferences** gives you direct access to **investors, developers**, and **city officials** who can help you navigate deals and opportunities.

Meetup.com & Local REIA Groups: Real Estate Investor Associations (REIAs) are a great place to find **investors, wholesalers**, and **potential buyers** for your land deals. These groups are filled with people **actively doing deals** and can open doors to opportunities you **wouldn't find elsewhere**.

Community Meetings & Development Trackers: Many cities have **development trackers** that list upcoming projects, rezonings, and land use changes. **Attending these meetings** is a goldmine for land investors because:

- You get **insight into where the market is heading** before everyone else.
- You meet **city officials and council members** who control zoning approvals.
- You connect with **builders and developers** who may need land for their next project.

Building Relationships with City Officials: Whether you realize it or not, city officials **hold the keys to development**. Getting to know **planners, zoning officials**, and **council members** helps you stay ahead of zoning changes, learn about upcoming city projects, and **build credibility when submitting entitlement applications**.

Networking with Brokers, Agents, and Wholesalers

Building strong relationships with **brokers, real estate agents**, and **wholesalers** is a **key strategy** in finding **off-market land deals** and getting access to opportunities before they hit the open market. Many agents, especially those **who don't specialize in land**, may not fully understand how to price or market vacant lots, which can work to your advantage. By connecting with land-focused agents, you can **gain early insight into new listings**, get notified about **price drops**, and **build rapport** so they bring you deals first.

Wholesalers, on the other hand, often secure off-market land at deep discounts. Many **wholesalers don't specialize in land**, so they might pass on deals that could be great entitlement plays. If you position yourself as a **go-to land buyer**, they will bring deals to you **before blasting them to their lists**.

Brokers who focus on **commercial or development land** can also be invaluable. These brokers work directly with large **landowners, builders, and developers** and can **bring you deals that fit your criteria**. Attending networking events, scheduling **one-on-one meetings**, and consistently following up with agents and wholesalers ensures that you are always **top-of-mind** when a new land opportunity arises.

Following up with **brokers and wholesalers** is just as crucial as following up with sellers. These professionals **constantly come across deals**, and if you're not staying in touch, you'll likely miss out on prime opportunities. Many investors make the mistake of reaching out once and expecting deals to come their way. Instead, **incorporate follow-ups into your marketing plan**—check in with brokers and wholesalers regularly, remind them of your buying criteria, and build relationships so that you're the first call when a good deal crosses their desk. A simple **monthly or biweekly check-in** via text, email, coffee (which works best), or a quick phone call can ensure you stay on their radar and get priority access to new opportunities.

Successful land acquisition is more than just finding properties—it's about having the right systems, strategies, and follow-up processes in place. Whether you're using direct mail, cold calling, networking, or digital marketing, consistency is key. Every method we've covered— GIS mapping, list building, social media outreach, and strategic follow-ups—plays a role in positioning you ahead of the competition. It's also crucial to have tools that keep you organized, like Asana, Monday, or Air table.

But finding the deal is only half the battle. Once you identify a piece of land, you need to determine whether it's actually worth pursuing. This is where market research and due diligence come in.

Now that we've successfully identified and engaged with sellers, it's time to shift our focus to the next critical step—**finding the right buyers and turning these deals into profits.** Having land under contract is great, but it doesn't mean much until you have an exit strategy in place. Whether you're wholesaling, flipping entitlements, or developing the land yourself, building a **strong, well-targeted buyer list** is essential for maximizing value and ensuring smooth transactions. The best buyers are often **right in the neighborhood**, actively working on projects or looking for their next deal. Now, let's dive into how to **strategically build a buyer network** that ensures you can close quickly and profitably.

BUILDING A BUYERS LIST

A well-curated buyers list is a crucial tool in land development and entitlement. Knowing who your end buyers are will help you tailor your acquisitions, negotiate better deals, and exit faster with maximum profit. Unlike traditional real estate, where buyers are mostly home-owners, land buyers are typically builders, developers, or investors looking for their next project. To build an effective buyers list, you need to categorize your buyers, research who is actively purchasing land, and establish strong relationships with them. Here are the main types of buyers you should focus on:

1. *Small Infill Investors*

These are local investors and small builders who specialize in developing single-family homes or duplexes in established neighborhoods. They are great for smaller parcels or subdivided lots and usually look for lots that already have infrastructure like water, sewer, and road access. Many of them are mom-and-pop builders or house flippers looking to build and sell new homes.

Where to Find Them:

- Look at tax records for LLCs or small corporations repeatedly buying lots in your target neighborhoods.
- Check MLS for new construction sales and see who the owner or developer is.
- Attend local real estate meetups or investor groups.
- Connect with agents who specialize in new construction sales and see if they represent builders looking for more land.

2. *Higher-End Custom Builders*

These builders focus on high-end custom homes in desirable neighborhoods. They look for premium lots with unique features—lake views, larger lot sizes, or locations in affluent communities. These buyers typically require more due diligence and may need additional zoning or permitting flexibility to meet their design needs.

Where to Find Them:

- Look for high-end new construction listings and check public records to see who the builder is.
- Reach out to luxury real estate agents who work with high-end custom home clients.
- Attend high-end home tours and builder showcases in your area.

- Search online directories or builder association lists for custom home builders in your market.

3. *Regional & National Home Builders*

Larger builders like **Lennar, DR Horton, Pulte, Toll Brothers**, and others are constantly looking for land in growth markets. These buyers prefer larger parcels they can subdivide and develop into entire communities. However, they also buy entitled or partially entitled land, making them a great option for developers looking to exit without doing the full development work.

Where to Find Them:

- Search public land records for large-scale purchases by major builders.
- Look at subdivision filings and plat approvals on city planning websites.
- Contact land acquisition managers at these companies (many have dedicated teams for sourcing land deals).
- Attend industry events like **National Association of Home Builders (NAHB) conferences** or regional builder expos.
- Drive through new developments and look for builder signs— these often have direct contact information.

4. *Investor-Sellers as Buyers*

One of the most overlooked sources of buyers is **other land investors and sellers** in your market. Many landowners who sell to developers are actually developers themselves. The best buyers are often already in the neighborhood, buying and selling land regularly.

Where to Find Them:

- Search **tax records** for LLCs or individual investors who own multiple lots or recently sold land.

- Look at **recent sales on the MLS** and see who purchased vacant lots. If they bought one, they might want more.
- **Call local builders** and ask if they are looking for additional land in the same area.
- Reach out to agents with **active new construction listings**—they may know builders who need more lots.
- Look at **development permits** filed in city planning records—developers who are actively building will likely need more land.

5. *Builders Working on Assemblages*

Some builders aren't just looking for single lots—they are trying to assemble multiple parcels to create a larger project. If you know a builder is in the process of assembling land in an area, they could be a perfect buyer for land nearby.

Where to Find Them:

- Look at **rezoning applications** and **development trackers** in city planning portals and attend city council meetings.
- Watch for **bulk purchases** in tax records—if someone bought multiple parcels in a short time, they are likely assembling land.
- Talk to local **zoning attorneys and engineers**—they often work with builders on land assemblages and can connect you to buyers.

MANAGING YOUR BUYERS LIST

Once you've identified and categorized your buyers, you need a system to **track and manage** them. Your buyers list should include:

- **Name & Company:** Who they are and what company they operate under.

- **Contact Information:** Phone number, email, and preferred method of contact.
- **Buying Criteria:** What types of land they buy (size, zoning, price range).
- **Geographic Preference:** Which neighborhoods, cities, or regions they focus on.
- **Recent Deals:** Have they closed on land recently? If so, what type?
- **Notes on Conversations:** Any specific details they've mentioned, such as financing needs or timeline preferences.

Using a **CRM (Customer Relationship Management)** system like **Podio**, **REI BlackBook**, or **Salesforce** can help you keep track of your buyers and follow up with them regularly. Many land deals require **persistence** and **timing**, and just because a builder isn't ready today doesn't mean they won't be six months from now.

BEING INTENTIONAL ABOUT YOUR NETWORKING

It's not enough to simply meet people at events or add their number to your phone. You need to **follow up**, **set up meetings**, and **deepen those relationships**. Here are a few best practices:

- **Set Up Coffee Meetings & Zoom Calls:** A quick phone call is fine, but nothing replaces a face-to-face sit-down or a dedicated Zoom call. The best deals come from **strong relationships**, and that happens when you take time to build trust.
- **Ask the Right Questions:** Instead of just introducing yourself, ask **who they know** and **who they work with**. Many of the best connections come through **referrals**. Some great questions include:
  - Who do you typically work with on your deals?
  - Do you know any investors, developers, or lenders looking for deals?

- ○ Who are your go-to contractors or subcontractors?
- ○ What challenges do you see in the market right now?
- **Follow Up & Stay in Touch:** Networking isn't a one-time event. You should be keeping track of who you meet and **following up regularly**. A simple **check-in message, market update**, or **invitation to grab coffee again** can keep you on their radar.
- **Keep a Networking Database:** Use **Excel, Google Sheets**, or a **CRM system** (like Podio or REI BlackBook) to log everyone you meet. This should include:
  - ○ Name
  - ○ Contact information
  - ○ Industry/expertise
  - ○ Who they work with
  - ○ Last interaction/date of last follow-up
  - ○ Notes on what they're looking for or specialize in

BUILDING A NETWORK OF CONTRACTORS & SUBCONTRACTORS

As a land developer or investor, **you don't want to be scrambling for contractors last minute**. The best way to avoid that is to **build relationships with experienced professionals early** so you have a go-to team ready when you need them.

Who You Need on Your Team

- **General Contractors (GCs):** They oversee the entire construction project and coordinate all subcontractors.
- **Subcontractors:** These include grading companies, utility installers, plumbers, electricians, framers, and roofers.
- **Surveyors:** They help with property boundaries, site plans, and subdivision approvals.
- **Civil Engineers:** They assist in site planning, stormwater drainage, and infrastructure planning.

- **Architects & Designers:** They are essential for building design, ensuring compliance with local codes, and maximizing space efficiency.

Where to Find Them:

- Ask **local builders** who they trust.
- Attend **local builder & developer meetups**.
- Look at **who pulled recent permits** for projects in your area (search in city permit databases).
- Visit **job sites** and talk directly to crews.
- Search contractor directories like **Angi**, **Houzz**, or the **NAHB**.

Vet Before Hiring: Always check **past work**, **reviews**, and **references** before bringing someone on board. A bad contractor can **derail a project**, **cost you thousands**, and **waste valuable time**.

FINDING & NURTURING LENDER RELATIONSHIPS

Unless you are paying **all cash for every deal**, you will need **lenders** and **financial partners** to help fund your acquisitions and developments.

Types of Lenders to Network With

- **Private Lenders:** Individuals who invest in real estate loans for passive returns.
- **Hard Money Lenders:** Provide short-term, asset-backed loans for acquisitions and construction.
- **Commercial Banks & Credit Unions:** For long-term development loans or refinancing.
- **Bridge Lenders:** Short-term financing options to bridge the gap between acquisition and long-term financing.
- **Institutional Investors & Equity Partners:** Larger funding sources for major projects.

How to Find Lenders:

- Attend **real estate investment and lending events**.
- Network with **other investors**—ask who they use for financing.
- Visit **local banks and credit unions**—many have loan programs specific to land and development.
- Search for **private lenders** on **LinkedIn** and **investment forums**.

Building Strong Lender Relationships:

- **Don't wait until you need financing!** Start conversations early.
- **Get pre-approved or at least gauge what financing options are available.**
- **Ask for their lending criteria.** Every lender has different requirements.
- **Keep them updated on your deals!** Even if you don't need financing today, sending updates on your projects keeps you top-of-mind.

EXPANDING YOUR NETWORK THROUGH OTHERS

Your success in land development will **largely depend on who you know**. When you meet someone new, don't just see what they can do for you—**see how you can help them** and how you can connect them to others in your network.

Leverage Referrals: If you meet an investor, ask them **who they trust** for deals, financing, and construction. If you meet a builder, ask who their go-to land acquisition people are.

Attend Local Networking Events: Builders, developers, lenders, and investors all have **networking events**, **industry conferences**, and **local**

meetups. Attending these is a great way to **organically build relationships** that will be useful in the future.

Be the Connector: Introduce people in your network to each other. If a lender needs deals and you know an investor with good deals, connect them! When you **help others first**, you naturally become the person they want to work with.

Host Your Own Meetups or Zoom Calls: If you don't see enough networking opportunities, **create your own!** A simple **monthly Zoom call** for investors, builders, and contractors to share market updates and deals can quickly **position you as an industry leader**.

Use Social Media to Your Advantage:

- **Create Facebook Groups** focused on land, development, and investments in your market.
- **Join LinkedIn Groups** where developers and builders connect.
- **Engage on Instagram & Twitter** with industry professionals.
- **Post valuable content**—market trends, deals, or zoning insights—to attract potential partners.

FINAL THOUGHTS: YOUR NETWORK = YOUR NET WORTH

At the end of the day, **real estate and land development is a relationship business**. The more people you know and the **stronger your connections**, the **more deals you'll close, the better financing you'll get**, and **the smoother your projects will run**. So don't hesitate to **set up coffee meetings, get on Zoom calls, attend networking events, find out who people know and get referrals**, and most importantly, **keep detailed records of your contacts and follow up!**

Before we get to the next section take a look at this example marketing plan you could use as your template:

EXAMPLE MARKETING PLAN

Daily Tasks

- Cold Calling & Texting (50–100 calls/day)
- Follow-Ups with Sellers, Agents & Wholesalers
- Lead Management & CRM Updates
- Social Media Engagement & Online Networking

Weekly Tasks

- Review & Optimize Cold Call & Texting Campaigns
- Begin New Direct Mail Campaigns
- Refine Social Media & Digital Marketing Strategy
- Check In With Brokers, Agents & Wholesalers
- Monitor Paid Ad Performance

Monthly Tasks

- New List Pull & Data Refinement
- Direct Mail Drop
- Agent, Broker, & Wholesaler Follow-Ups
- Marketing Review & Strategy Adjustment

Quarterly Tasks

- Deep Review of Marketing ROI
- Adjust Marketing Spend Based on Performance
- Optimize Skip Tracing & Data Strategy
- Refine Scripts & Cold Call Training
- Attend at Least One Major Networking Event

Yearly Tasks

- **Complete Annual Performance Review**
- **Adjust Annual Marketing Budget**
- **Expand Professional Network**
- **Implement New Technology & CRM Enhancements**
- **Consider Hiring Additional Team Members**

DEVELOPING LAND DEVELOPERS

"It is nice to have valid competition; it pushes you to do better."

GIANNI VERSACE

I think of land development as a largely untapped goldmine. Of course, people are doing it, but it has far more potential than is being realized right now. Throughout my own journey so far, the fact that so few people are doing it has worked in my favor, and it will in yours too. You might wonder, then, why I'd want to write a book about this and share what I've learned with more people who might later prove to be my competition.

The answer is that it will only help the people who are already interested anyway. There are still a great many people who will be put off by the risk involved, the amount of time it can take, and the upfront cost. We all need a bit of healthy competition, and this book will only help people who already *are* that competition. It will just give them the tools they need to increase their confidence and see success. Besides, there's a lot of land out there, so we can all succeed.

You are one of these people. You may not have tried your hand at this yet, or you may have minimal experience, but wherever you are on your journey, you're someone who's not afraid of a little risk and who has the financial security to make at least one upfront payment. So I want to help you. This is an exciting venture, and I want to share that with you. In fact, I'd like to share it with everyone in a similar situation —and this is your chance to help me.

If you'd be willing to spare just a few minutes of your time to leave your feedback online, you'll help other people who are genuinely interested in land development find it. A little competition is a good thing, and so is a solid network of people who know how this works.

By leaving a review of this book on Amazon, you'll be contributing to that community and helping someone else take a successful plunge into land development.

There are readers who are looking for this information already; your review will help them to find it quickly and easily, without falling for misinformation.

Thank you for your support. It's great to have you onboard!

Scan the QR code below.

CHAPTER 3
MARKET RESEARCH & DUE DILIGENCE—THE KEY TO PROFITABLE LAND DEALS

F inding a piece of land is just the first step—now, you need to determine whether it's actually a **good deal**. Too many investors jump into contracts without fully understanding zoning restrictions, infrastructure challenges, or market conditions, only to find out later that they've overpaid or bought something they can't develop.

This chapter is all about **mitigating those risks before you buy**. We'll cover how to research zoning codes, analyze local market trends, assess infrastructure availability, and conduct thorough due diligence. By mastering these skills, you'll be able to confidently identify land that has **real** development potential—while avoiding costly mistakes that could set you back months (or even years).

It's crucial to remember that while you're orchestrating the deal, you are not the engineer, attorney, surveyor, or contractor—your role is to assemble the right professionals and let them do their jobs. However, that doesn't mean you should operate blindly. You need to have a working knowledge of each role to ensure things are moving in the right direction. Understanding how to read bids, hiring contracts, surveys, and basic engineering principles will prevent costly mistakes and miscommunications. There will be times when city officials or even hired professionals make errors, misinterpret zoning codes, or overlook crucial details, and if you don't have a foundational under-standing of the process, you won't catch these issues before they become costly delays. Knowing how to review municipal codes, inter-pret surveys, and grasp essential engineering concepts allows you to ask the right questions and push things forward when necessary. And when things get legally complex, don't hesitate to bring in attorneys to review documents and ensure you're protected. Being informed doesn't mean you have to be the expert—it means you have enough knowledge to manage the experts effectively.

Let's get started!

When researching markets and conducting due diligence, **you need to start with the end in mind**. Your **exit strategy** will determine every-thing—how you structure the deal, what due diligence steps to take, and even how you finance the purchase. The key is to have **multiple exit strategies** in case your initial plan doesn't work out. Unlike other areas of real estate where dozens of strategies exist, land development typically has a handful of clear-cut options. However, **having more than one path forward** is essential to minimizing risk and ensuring profitability.

Another important thing to keep in mind is that exit strategies in land deals are rarely rigid—they often evolve as new opportunities arise during the contract period. What starts as a straightforward entitlement play might turn into a joint venture with a builder, or a land banking strategy could shift into a quick flip if the right buyer emerges. Some-times, a combination of strategies is the best path forward, and you have to be adaptable enough to pivot when necessary. Market condi-tions, zoning approvals, financing options, and buyer interest can all influence your final decision, making it critical to structure your contract with enough time and contingencies to explore multiple avenues. The ability to make high-level, fast decisions is key—whether it's recognizing an opportunity for higher density, selling entitled land instead of developing, or leveraging partnerships for a bigger exit. Your job is to stay ahead of the curve, assess new variables as they come up, and ensure your contract terms allow you the flexibility to capitalize on the best possible outcome.

COMMON EXIT STRATEGIES FOR LAND DEVELOPMENT

1. **Wholesale:**

The simplest and fastest way to exit a deal. You **contract the land at a discount** and **assign or double-close it** to a builder or investor for a profit.

- **Example:** You contract a lot for $500K and assign it to a builder for $550K, making a $50K wholesale fee.

2. **JV with a Builder or Investor:**

Instead of selling outright, you **partner with a builder or a capital investor** to develop the land together, sharing in the profits.

- **Example:** A builder provides the funding and construction expertise while you handle the entitlement process. Profits are split at an agreed percentage.

3. **Hold for Appreciation (Land Banking):**

Buying land in an area of future growth and **holding on to it** while its value increases. This requires patience and the ability to cover carrying costs.

- **Example:** You buy a ten-acre parcel near a growing city, hold it for five years, and sell it for three times the price as demand increases.

4. Sell Entitled Land:

Instead of holding long-term, you **entitle the land** by securing zoning approvals and allocated necessary infrastructure, then sell it to a builder. You want to be careful of going too far with it because if you go as far as permits, then you limit the number of buyers because you may have permitted against their regulations, therefore not leaving any room for the end buyer to do what they want.

- **Example:** You buy raw land for $1M, entitle it for fifty townhomes, and sell it to a developer for $3M.

5. Develop and Build Yourself:

The highest-profit but most complex strategy. This involves **buying land**, **entitling it**, and **constructing homes or commercial buildings**. It requires financing, construction experience, and management.

- **Example:** You entitle land for thirty units, hire a general contractor, and build homes to sell at full retail value.

MARKET RESEARCH: ANALYZING THE END PRODUCT AND FUTURE MARKET CONDITIONS

The key to successful land development is not just **finding a good deal** but **understanding what the end product will look like and how the market will respond** by the time the project is completed. Market conditions can shift dramatically over the course of a land deal, so before moving forward, you need to **analyze key factors** that will help predict whether your project will be viable in the future.

Understanding the Timeline & Market Forecasting

A land deal doesn't happen overnight. From **contract** to **permitting**, **infrastructure**, **vertical construction**, and **final sales**, the process can take years. That means you're not operating in the market **as it exists today**—you're operating in the market **that will exist when your product is complete**.

Before locking in a deal, ask yourself:

- **How long will it take to entitle the land and secure approvals?** (Rezoning, variances, site plan approvals).
- **How long will infrastructure take?** (Sewer, roads, stormwater, grading, utilities).
- **When will the actual construction begin and end?**
- **By the time the homes or buildings are ready to sell, what will the market look like?**

These answers will determine if the project is worth pursuing and what direction or directions to take.

Analyze Key Factors

One critical thing you need to do is **research the local market conditions** to ensure the numbers and project make sense. Here are some crucial things to focus on:

1. **Proposed Units in the Pipeline:** Look at city planning and development trackers or go visit the city planner to see **how many similar projects are being proposed and built in the area**. If the market is already flooded with inventory, your project may struggle.
2. **Current Market Demand & Price Trends:** Study **what's selling, how fast,** and **at what price.** Track sales trends over the last few years to identify any major shifts in demand.
3. **Major Developers in the Area:** Find out **who the dominant builders are, what they are building,** and **what their strategies are.** This can be done by:

- o **Talking to brokers & agents who specialize in land and new construction**
- o **Searching MLS for listings of homes in new developments**
- o **Looking at development trackers to see what's in permitting**

4. **Sewer Capacity & Infrastructure Availability:** You can't build if there's no **sewer capacity**, **water access**, or **road infrastructure**. Contact local municipalities to ensure there's available capacity **not just for existing projects but for your own as well.**

5. **Hidden Off-Market Development Deals:** Many of the largest builders don't list their properties on the MLS. Instead, they advertise their communities **through site plans, developer websites**, and **marketing materials. Find out what's happening behind the scenes** so you're not caught off guard by a flood of competing inventory.

6. **Sales Price Per Square Foot for Finished Homes:** Understanding **the price per square foot** of **finished homes in the area** helps determine:
   - o What the market is **willing to pay**
   - o What product types **(bed/bath count, size, finishes)** perform best
   - o Whether a deal is **worth pursuing**

7. **Competition & Amenities Analysis:** Look at **nearby communities** and track what amenities they offer: **pools, dog parks, clubhouses, trails**, etc. Builders **compete not just on price but on lifestyle offerings.** Make sure your development aligns with buyer expectations.

8. **Determining How Much Builders Paid for Their Land:** This is **one of the most important factors** in pricing your own land deal. Builders often **assemble multiple parcels** rather than buying single lots. If you can **track previous sales records**, you can **estimate their cost per pad** and **reverse-engineer pricing.**

Understanding Raw vs. Finished Pad Pricing

One of the biggest **mistakes new land investors make** is assuming that **land is just land**. However, there's a big difference between a **raw land price** and a **finished pad price**.

1. **Raw Land Price** = Before any infrastructure is added (sewer, roads, grading)
2. **Finished Pad Price** = Fully developed lot ready for vertical construction

A finished pad is always **significantly more expensive and valuable** than raw land because the cost of roads, utilities, and land prep has been factored in. Your job as an investor is to:

1. Estimate the finished pad price by studying what builders are paying
2. Estimate the cost of infrastructure to understand total development costs

Estimating Infrastructure Costs (Without Overanalyzing)

The key here is **to get rough estimates, NOT actual contractor bids**. Since **you're not the builder, you don't need detailed construction quotes**—you just need a general sense of costs so you don't overpay for land. Market research and due diligence start with **determining your end product** and working backward. By analyzing local trends, understanding infrastructure costs, and tracking what builders are doing, you can make informed decisions and avoid costly mistakes.

Infrastructure costs can vary widely, depending on:

1. Site topography (flat land vs. sloped land)
2. Distance to utilities (sewer, water, electric)
3. Road construction and stormwater requirements

Why Not Get Bids Upfront?

1. **Costs Change Over Time:** If you're not building for another two to three years, today's material and labor costs won't be accurate.
2. **Every Builder Has Different Requirements:** Some builders need luxury finishes, while others need budget-friendly solutions. The cost varies.

Instead, use **industry averages** and local builder insights to make rough projections. This is a great opportunity to call builders and get their insight, tell them you may have a potential deal for them, and find out what they are paying for raw and finished pads.

Now that we understand how to research the **market demand**, let's move into the **due diligence process**—where we verify zoning, infrastructure, title work, and other critical factors before making an offer.

DETERMINE HIGHEST AND BEST USE

Highest and best use refers to the **most profitable** and **legally permissible** use of a property. It considers physical, legal, and financial factors to determine what type of development would generate the greatest value.

For example, a vacant ten-acre parcel in an urban area might be zoned for single-family homes but could be **rezoned for multifamily apartments**—a change that could significantly increase its value. Conversely, a piece of land in a rural area might be better suited for agricultural use than high-density development due to lack of infrastructure.

By understanding HABU, investors and developers can make **informed decisions**, ensuring they maximize profits while avoiding costly mistakes.

4 Most Common Tests of Highest and Best Use

Consider the following **key factors**:

1. **Legally Permissible:** What does the zoning code allow? Are there restrictions such as overlays, HOAs, or environmental protections?
2. **Physically Possible:** Does the land have the right topography, soil conditions, and infrastructure for the intended development?
3. **Financially Feasible:** Will the proposed development generate enough revenue to justify the cost of acquisition, entitlements, and construction?
4. **Maximum Density:** Among the feasible uses, which one provides the **greatest ROI**?

Example: A five-acre plot of land near a major highway could be developed into **single-family homes**, a **shopping center**, or a **storage facility**. While all may be physically possible, zoning might restrict it to commercial use. If a **storage facility generates the highest revenue** based on demand, that would be its HABU. One thing to remember is that maximum density may not always be highest and best use.

Zoning and Land Use Regulations

1. **Zoning Classifications & Land Use Designations**

- **Zoning Code & Districts:** Verify the current zoning and determine what is allowed under that designation.
- **Future Land Use Plans:** Check the city or county's comprehensive plan to see if there are planned zoning changes that could affect the property.
- **Mixed-Use or Overlay Districts:** Some areas have special zoning overlays that allow for different uses or additional development incentives.

2. **Rezoning & Variances**

- **Rezoning Process:** If the current zoning does not allow your intended use, research the rezoning process, requirements, and likelihood of approval.
- **Variance Requests:** Some developments may require zoning variances to adjust setbacks, density, height, or use restrictions.
- **Community & Political Support:** Even if zoning changes are legally possible, local opposition or city council members can influence approvals.

3. **Density & Development Restrictions**

- **Maximum Density Limits:** Research how many units or structures are allowed per acre under the zoning code.
- **Floor Area Ratio (FAR):** Check FAR limits to see how much building square footage you can construct relative to the lot size.
- **Lot Coverage & Open Space Requirements:** Some zoning codes require a certain percentage of the property to remain as green space.

4. **Setbacks, Height Limits & Building Envelopes**

- **Front, Side, & Rear Setbacks:** Ensure the required distance from property lines doesn't limit your intended development.
- **Height Restrictions:** Some zoning codes have strict limits on building height, especially near residential neighborhoods or historic districts.
- **Building Envelopes:** Review diagrams or legal descriptions to understand the maximum buildable area within the zoning regulations.

5. **Parking & Traffic Requirements**

- **On-Site Parking Minimums:** Determine the required number of parking spaces per unit, commercial square footage, or use type.
- **Traffic Impact Studies:** Some municipalities require traffic studies to analyze how the development will affect roadways and intersections.
- **Ingress/Egress & Loading Zones:** Commercial and mixed-use projects may need specific driveways, turn lanes, or loading areas.

6. **Environmental & Conservation Restrictions**

- **Floodplain & Wetland Protections:** Properties in flood-prone areas may require additional approvals, engineering solutions, or restrictions on development.
- **Stormwater Management Requirements:** Certain zoning areas require stormwater mitigation systems, retention ponds, or green infrastructure.
- **Tree Preservation & Landscape Buffers:** Many municipalities have tree ordinances that require preservation or replanting of trees.

7. **Special Use Permits & Conditional Uses**

- **Special Use or Conditional Use Permits:** Some zoning districts allow certain types of development only with special approvals.
- **Planned Unit Developments (PUDs):** Some developments require a PUD agreement, which can be more flexible but requires more oversight and approvals.
- **Short-Term Rental & Use Restrictions:** Some areas have strict rules against Airbnb, vacation rentals, or commercial uses in residential zones.

8. **Utility & Infrastructure Requirements**

- **Water & Sewer Availability:** Confirm whether the zoning allows for the required utility connections or if additional infrastructure improvements will be necessary.
- **Septic vs. Sewer Requirements:** If public sewer isn't available, research septic system approvals and minimum lot sizes for septic use.
- **Electrical & Gas Service:** Verify if new utility lines need to be extended and if zoning codes allow overhead or underground utilities.

9. **Impact Fees & Development Costs**

- **Impact Fees:** Some municipalities charge fees for new developments to offset costs for roads, schools, or public services.
- **Permit & Connection Fees:** Research costs for building permits, water/sewer tap fees, and other regulatory expenses.
- **Affordable Housing or Community Contributions:** Some zoning laws require a percentage of units to be affordable housing or for the developer to contribute to community projects.

10. **Historic Districts & Design Guidelines**

- **Historic Overlay Districts:** If the property is within a historic district, special restrictions may apply to new construction or renovations.
- **Architectural Review Boards:** Some areas require design approval for new developments to ensure compatibility with the surrounding neighborhood.
- **Neighborhood or HOA Restrictions:** Check for private restrictions or design guidelines that may affect the project.

11. Annexation & Jurisdictional Issues

- **City vs. County Regulations:** Some properties sit in county jurisdictions but may be annexed into city limits, changing the applicable zoning and permitting rules.
- **Annexation Agreements:** If the property is in an area where annexation is likely, research what zoning and service requirements will apply.
- **Multiple Jurisdictions:** Some properties may be affected by both city and county rules, requiring coordination between different agencies.

12. Development Moratoriums & Political Climate

- **Temporary Building Moratoriums:** Some cities impose temporary restrictions on new developments due to infrastructure concerns or zoning updates.
- **Local Political & Community Sentiment:** Even if a project meets zoning requirements, strong community opposition can delay or stop approval processes.
- **State vs. Local Development Laws:** Some states have laws that override local zoning rules, such as density bonus programs or statewide housing mandates.

Economic Feasibility Analysis

1. Acquisition & Holding Costs

- **Purchase Price:** Ensure that the land cost aligns with market comparables and your projected development margins.
- **Earnest Money & Deposits:** Consider the upfront costs required to secure the property under contract.
- **Property Taxes:** Calculate annual property taxes and assess whether the holding period justifies the costs.

- **Carrying Costs:** Account for insurance, maintenance, and any HOA or special district fees.

2. **Entitlement & Pre-Development Costs**

- **Zoning & Entitlement Fees:** Research costs for rezoning applications, variances, and special permits.
- **Surveying & Engineering Studies:** Factor in costs for boundary surveys, soil tests, geotechnical reports, and stormwater management plans.
- **Architectural & Design Fees:** Include expenses for master planning, building layouts, and site design.
- **Legal & Consulting Fees:** Account for attorney fees for contract negotiation, zoning approvals, and compliance with local regulations.

3. **Infrastructure & Site Development Costs**

- **Road Construction & Access Improvements:** Determine if new roads or widening of existing roads are required.
- **Sewer & Water Line Extensions:** Research if utility connections exist or if you'll need to install costly infrastructure.
- **Stormwater Drainage & Retention Ponds:** Estimate costs for detention ponds, culverts, and drainage improvements.
- **Environmental Mitigation Costs:** Budget for any environmental cleanup, tree mitigation fees, or flood zone requirements.

4. **Construction & Vertical Development Costs**

- **Per Unit or Per Square Foot Costs:** Determine the estimated cost per unit or square foot based on comparable projects.
- **Material & Labor Costs:** Consider current market rates for materials, workforce availability, and inflationary risks.

- **Permitting & Impact Fees:** Research local impact fees, building permits, and any additional regulatory costs.
- **Construction Financing Costs:** Calculate loan origination fees, interest rates, and construction draw schedules.

5. **Market Demand & Sales Price Projections**

- **Comparable Sales Data:** Analyze recent sales of similar properties, including price per square foot and price per lot.
- **Absorption Rates:** Determine how quickly similar properties are selling or leasing in the market.
- **Product Type & Consumer Demand:** Assess whether the proposed product type (residential, commercial, mixed-use) is aligned with market needs.
- **Buyer & Tenant Preferences:** Research whether certain home features, lot sizes, or commercial layouts are preferred by buyers.

6. **Exit Strategies & Revenue Projections**

- **Wholesale Land Sales:** If flipping the land before development, determine potential profit margins.
- **Entitled Land Sales:** If selling after entitlements, estimate the added value and demand for shovel-ready sites.
- **Develop & Sell vs. Build-to-Rent:** Weigh the financial benefits of selling units individually versus holding and renting.
- **Subdivision & Lot Sales:** Determine if subdividing the land and selling finished lots provides a higher return.

7. **Financing & Capital Stack**

- **Debt vs. Equity Financing:** Assess whether project funding will come from loans, investor capital, or self-funding.

- **Loan Terms & Interest Rates:** Compare lender terms, including debt service coverage ratios and loan-to-cost requirements.
- **Equity Partner Expectations:** If using investors, structure agreements to align with profit-sharing and risk tolerance.
- **Bridge & Mezzanine Financing:** Consider short-term financing solutions if construction funding takes longer than expected.

8. **Risk Assessment & Sensitivity Analysis**

- **Cost Overruns & Budget Contingencies:** Include a 10–20 percent contingency in the budget for unexpected costs.
- **Market Downturn Risks:** Consider the impact of economic slowdowns or rising interest rates on project viability.
- **Regulatory & Political Risks:** Research any proposed changes in zoning laws, tax incentives, or building codes that could affect profitability.
- **Exit Timeline & Liquidity Risks:** Ensure that the project timeline aligns with market conditions and investor expectations.

9. **Break-Even Analysis & ROI Calculations**

- **Cap Rate & Yield Expectations:** For income-producing properties, determine the projected cap rate and net operating income (NOI).
- **Return on Investment (ROI):** Calculate total project costs versus expected revenue to assess profitability.
- **Break-Even Sales Volume:** Determine how many units or lots need to sell before the project becomes profitable.
- **Internal Rate of Return (IRR):** Use IRR to compare the project's potential return against alternative investments.

10. **Alternative Revenue Streams & Value-Add Opportunities**

- **Short-Term Leases or Interim Uses:** If development is delayed, consider temporary income from parking, billboards, or agricultural leases.
- **Tax Incentives & Grants:** Look for historic preservation credits, affordable housing incentives, or opportunity zone benefits.
- **Seller Financing or Creative Deal Structuring:** Explore options like seller carrybacks, joint ventures, or option agreements to reduce capital needs.

11. **Competition & Market Positioning**

- **Competing Developments in the Pipeline:** Research upcoming projects that may impact pricing and demand.
- **Major Developers & Builders:** Identify key players in the market and their current acquisition strategies.
- **Unique Selling Proposition (USP):** Define what makes your project stand out from the competition.
- **Marketing & Pre-Sales Strategy:** Ensure there is a plan to attract buyers or tenants well before project completion.

12. **Long-Term Holding vs. Immediate Sale**

- **Holding for Appreciation:** Determine whether it's financially beneficial to hold the land for long-term appreciation.
- **Cash Flow Considerations:** If holding, analyze rental income potential and operational costs.
- **1031 Exchange & Tax Planning:** Explore tax-deferred exchange options if reinvesting proceeds into another property.
- **Legacy Planning & Exit Timelines:** If passing the property to future generations, consider estate planning and transfer tax implications.

Physical Site Conditions

1. **Roads & Access**

- **Existing Road Frontage:** Ensure the property has sufficient road frontage for the intended development. Limited access could require costly easements or road construction.
- **Road Width & Condition:** Check if the existing roads are wide enough to meet city or county requirements and whether they need repaving or expansion.
- **Traffic Volume & Flow:** Consider how busy the road is and if additional traffic studies will be needed. Certain roads may have restrictions for commercial traffic or require deceleration lanes.
- **Ingress & Egress:** Confirm whether there are existing curb cuts or driveways. If not, approvals may be needed for access points.
- **Connectivity to Major Roads:** Proximity to highways, main roads, and interchanges can impact desirability and zoning.
- **Planned Road Expansions:** Check city or county transportation plans to see if any road widening or changes could impact the property.

2. **Drainage & Stormwater Management**

- **Storm Drains & Retention Ponds:** Determine if the property has existing stormwater drainage or if you'll need to construct detention or retention ponds.
- **Flood Zones & Wetlands:** Review FEMA flood maps and wetland delineations. Development in flood-prone areas requires mitigation and additional approvals.
- **Topography & Water Runoff:** Identify slopes and natural drainage patterns to determine whether grading or retention structures will be needed.

- **Stormwater Easements & Infrastructure:** Check if stormwater drainage from neighboring properties flows onto your site, which could affect development plans.

3. **Utilities & Infrastructure**

- **Water & Sewer Availability:** Verify if public water and sewer lines are on-site or nearby. If not, alternative solutions like well water or septic systems may be needed.
- **Sewer Capacity & Connection Points:** Even if sewer lines exist, capacity may be an issue. Confirm with local utility departments whether additional infrastructure is needed.
- **Electrical & Gas Services:** Check utility maps to see if electricity and gas lines are readily available or if new connections will be required.

4. **Grading & Soil Conditions**

- **Soil Composition:** Conduct geotechnical testing to identify potential issues such as clay, rock, or unstable soil that could impact construction costs.
- **Elevation & Slopes:** Excessive grading may be required if the land is too steep, leading to higher costs for cut-and-fill operations.
- **Retaining Walls & Erosion Control:** Steep slopes may require retaining walls or erosion control measures to comply with local regulations.

5. **Easements & Right of-Way**

- **Utility Easements:** Determine if existing water, sewer, or power lines run through the property and if they could impact development plans.
- **Public Right-of-Way:** Check if any part of the property is reserved for public infrastructure expansion.

- **Private Access Easements:** Some properties may have shared driveways or access agreements that could affect how the land can be used.

6. **Environmental Factors**

- **Tree & Vegetation Protection:** Some municipalities have tree ordinances requiring preservation or replacement of removed trees.
- **Endangered Species & Habitat Restrictions:** Certain areas may have protected wildlife habitats that limit development potential.
- **Hazardous Waste or Contamination:** Conduct an environmental site assessment (ESA) to ensure there are no hazardous materials on the property.
- **Check for Wetlands & Flood Zones:** Identify environmental concerns that may limit development.
- **Evaluate Topography & Soil Conditions:** Determine whether grading or special construction techniques are needed. Always check for sinkholes and rock.
- **Consider Elevation & Views:** Proximity to major cities and amenities can affect premium pricing.
- **Road Access & Traffic Flow:** Ensure storm drains, traffic calming or control mitigation, connectivity to major roads, and future planned expansion are accounted for.
- **Know Your Market:** Every market is different, so be sure to learn the parameters of your specific location. A good rule of thumb is deducting **20 percent for roads** before calculating unit density.
- **Utility Availability:** Assess availability for water, sewer, gas, and electric, including water runoff and drainage.
- **Tree Density & Regulations:** Research local tree preservation laws and any arborist requirements that may apply.

Market Demand and Development Trends

- **Current Sales Trends:** Analyze what types of properties (single-family homes, multifamily units, mixed-use developments) are selling fastest in your target market. Study recent sales comps and absorption rates to understand demand.
- **Job Growth & Employment Centers:** Evaluate the local job market. Areas with strong employment growth and major employers nearby typically have more sustainable housing demand.
- **Population Growth & Migration Patterns:** Look at census data, state migration reports, and local demographic studies to see if people are moving into or out of the area.
- **Nearby Developments:** Monitor major projects such as new highways, transit stations, shopping centers, and hospitals, which can significantly impact land values.
- **Rental vs. Ownership Trends:** Determine whether buyers in the area prefer to rent or own, as this will influence whether you should focus on build-to-rent communities or for-sale housing.
- **New Construction Inventory & Pipeline:** Research how many new housing units are being built and whether the supply is keeping up with demand.
- **Price Trends & Affordability:** Look at median home prices, appreciation rates, and affordability indexes to understand market cycles and potential pricing ceilings.
- **Zoning & Policy Changes:** Stay informed on upcoming zoning adjustments or municipal incentives that may impact what can be developed on the land.
- **Builder & Developer Activity:** Identify which builders are active in the market and whether they are looking for more land to acquire.
- **Financing Conditions & Interest Rates:** Keep track of mortgage rates and lending conditions, as they directly impact buyer affordability and investor activity.

- **Housing Preferences & Design Trends:** Study consumer preferences, including home sizes, energy efficiency features, and community amenities that add value in the area.

Different Types of Projects Your Land Can Create

Residential and Multifamily Development

- **Infill for Townhome Projects:** A small vacant lot in an urban or suburban area may be best suited for a townhome community rather than a single-family home, maximizing density and value.
- **Single-Family Detached Subdivisions:** A ten-acre or larger property in a suburban or exurban area may be most profitable as a single-family home subdivision with varying lot sizes and amenities.
- **Luxury Single-Family Homes:** A prime location in an upscale neighborhood or scenic area (such as waterfront or hillside properties) may be best suited for high-end custom homes with premium amenities.
- **Garden Apartments:** A low-density, 2–3-story apartment community with surface parking may be ideal for suburban areas with moderate rental demand.
- **Podium Apartments:** A mid-rise apartment building with a structured parking podium could be best suited for an urban corridor where land values are higher.
- **Wrap Apartments:** A larger apartment development that wraps around a parking garage may be viable in a high-density urban or mixed-use district.
- **High-Rise Apartments:** A prime location in a major metro area could support a luxury high-rise residential building catering to high-income renters.
- **Build-to-Rent (BTR) Communities:** A suburban or exurban location may be best suited for a single-family rental

community offering detached homes with rental-style management.

Commercial & Mixed-Use Development

- **Retail Development:** A vacant parcel along a major highway may be best suited for a shopping center, gas station, or fast-food restaurant.
- **Office Space:** A well-located lot near a transit hub may be ideal for a medical office, co-working space, or corporate building.
- **Mixed-Use:** In a growing downtown, a residential-over-retail development could be maximally productive.
- **Hospitality & Lodging:** A site near a major highway, convention center, or airport may be best suited for a hotel or extended-stay property.
- **Self-Storage Facilities:** A high-traffic suburban or urban corridor with limited storage options could be an ideal location for a storage facility.
- **Entertainment & Lifestyle:** Land near a tourist attraction or high-growth area may be suited for a movie theater, sports complex, or event venue.

Industrial & Logistics

- **Warehouses & Distribution Centers:** A large, undeveloped lot near an airport or highway interchange might be ideal for a warehouse or distribution hub.
- **Fulfillment Centers:** In an area with a growing e-commerce industry, a fulfillment center may be the best use.
- **Flex Industrial Space:** A site with good access and zoning may be ideal for small-bay industrial spaces catering to contractors, small businesses, and light manufacturing.

Agricultural & Recreational

- **Agriculture & Farming:** A large tract of land in a rural area might be best suited for agriculture, conservation, or hunting leases.
- **Eco-Tourism & Outdoor Recreation:** A site near a national park, forest, or scenic area could be developed into a glamping resort, RV park, or adventure tourism facility.
- **Lakefront or Resort Communities:** A lakefront or coastal property could be turned into a vacation rental community, marina, or luxury second-home development.

Now that we've explored the highest and best uses for different types of land, it's time to dig deeper into the due diligence process using mapping tools, GIS systems, and municipal data. Understanding how to analyze zoning maps, floodplains, topography, infrastructure availability, and future development plans is essential to making informed investment decisions. Every city and county has different layers of information that can impact feasibility, from sewer capacity to road expansion plans. Proper due diligence ensures that the development you envision is not only legally permissible but also financially viable. In the next section, we'll break down how to navigate GIS systems, municipal data sources, and mapping tools to uncover critical details that could make or break your deal.

USING GIS & MUNICIPAL DATA FOR DUE DILIGENCE

Most cities and counties provide access to various mapping tools and Geographic Information System (GIS) data, which are crucial for land development due diligence. However, the availability, format, and level of detail in these maps can vary widely depending on the municipality. Some cities have highly sophisticated GIS platforms with multiple interactive layers, while smaller counties may only provide static PDF maps or require requests through planning departments. To find them, it's best to just google them. Here are some examples.

- **GIS Maps:** A digital mapping tool provided by cities and counties that overlays multiple data points, including zoning, parcel ownership, utilities, environmental constraints, and more. Essential for a broad view of site feasibility.
- **Zoning Maps:** Show current zoning classifications, land use restrictions, density limits, and permitted uses for specific parcels. Crucial for determining what can be built by right and what may require rezoning.
- **Future Land Use Maps (FLU):** Outline the municipality's long-term vision for growth and development, indicating areas planned for commercial, residential, industrial, or mixed-use expansion. Help assess the likelihood of rezoning approvals.
- **Comprehensive Plan Maps:** A policy document used by local governments that outlines long-term growth strategies, infrastructure expansion, transportation corridors, and urban development priorities.
- **Floodplain & FEMA Flood Maps:** Identify flood-prone areas and flood zones, determining whether the land will require costly mitigation such as raised foundations, retention ponds, or flood insurance.
- **Topographic (Contour) Maps:** Show elevation changes, slope grades, and natural land formations. Steep slopes may require excessive grading, while flat land may indicate poor drainage issues.
- **Soil & Geotechnical Maps:** Identify different soil types, drainage capabilities, and load-bearing capacities. Useful for assessing construction feasibility and foundation requirements.
- **Environmental & Wetlands Maps:** Display environmentally sensitive areas, wetlands, conservation overlays, protected species habitats, and hazardous waste sites. Essential for compliance with environmental regulations.
- **Stormwater & Drainage Maps:** Show existing storm drains, retention/detention ponds, watershed boundaries, and natural

water runoff patterns. Help determine whether additional stormwater infrastructure will be required.

- **Sewer & Water Infrastructure Maps:** Display existing sewer and water lines, pipe capacities, and connection points. Essential for understanding whether municipal services are available or if alternative solutions like septic systems or well water are needed.
- **Transportation & Roadway Expansion Maps:** Outline planned road widening, new highways, transit-oriented developments, and other transportation projects that could impact site access and value.
- **Utility Easement & Right-of-Way Maps:** Identify legal access for utility companies, road expansions, or other infrastructure improvements that may impact site usability or require relocation costs.
- **School District & Attendance Zone Maps:** Important for residential developments as they influence property values and buyer demand.
- **Parcel Ownership & Tax Assessment Maps:** Display ownership records, property boundaries, tax values, and recent sales history. Useful for evaluating property values and negotiating acquisitions.
- **Opportunity Zone & Economic Incentive Maps:** Highlight areas designated for tax benefits, economic development incentives, or investment-friendly policies that can increase project profitability.

UNDERSTANDING THE DIFFERENCE BETWEEN CITY AND COUNTY JURISDICTIONS

Understanding the distinction between city and county jurisdictions is crucial when navigating land development regulations, as each operates within its own governance framework and regulatory scope. Counties serve as administrative subdivisions of a state, encompassing multiple municipalities, cities, towns, or unincorporated areas. They typically manage services and regulations that span broader regions, such as public health, safety, and infrastructure in unincorporated areas —those not governed by a city or town. In contrast, cities are incorporated entities with their own local governments, possessing authority over municipal services, zoning, and land use within their defined boundaries. This means that properties located within city limits are subject to city ordinances and regulations, while those in unincorporated areas fall under county jurisdiction.

Given this structure, it's essential to identify the appropriate governing body for your specific area to ensure compliance with relevant regulations. For instance, if your property is within the city limits of St. Petersburg, Florida, you would refer to the city's Planning & Zoning Department for guidelines and permits. Conversely, if the property lies in an unincorporated part of Pinellas County, the county's Building and Development Review Services Department would be the correct point of contact. To determine the jurisdiction of a particular property, you can utilize tools like the "My Neighborhood Services" search provided by Pinellas County. Once the jurisdiction is established, reaching out to the respective department will provide you with specific information on zoning laws, land use regulations, and development guidelines pertinent to your project. Here's just a quick look.

- **City Land** is usually subject to **stricter zoning codes, development requirements, and infrastructure regulations**. Cities often have **comprehensive plans** outlining their vision for growth, and property owners may need to adhere to design overlays, zoning restrictions, and development policies.
- **County Land** typically has **fewer development restrictions** and more flexible zoning regulations. However, rural counties may lack essential infrastructure like **sewers, water**, and **public roads**, which can significantly impact development costs.
- **Annexation:** Some properties in county jurisdictions may be eligible for annexation into the city. This process allows

landowners to **tap into municipal services like sewerage and water**, but it may also impose additional zoning and tax burdens.

UNDERSTANDING MUNICIPAL CODES AND REGULATIONS

Understanding municipal codes and regulations is a crucial step in determining what can and cannot be done with a property. Each city and county has its own set of zoning laws, land use regulations, and building codes that govern development. While zoning maps give a broad overview, the municipal code provides the detailed rules on lot sizes, density, setbacks, permitted uses, overlay districts, and conditional approvals. It's essential to not only review zoning classifications but also examine subdivision regulations, environmental restrictions, parking requirements, and any design guidelines that could impact a project. Below is a quick overview of the key sections to focus on when reviewing municipal codes to ensure compliance and avoid costly surprises during the entitlement process.

Key Elements of Municode

- **Zoning Classifications:** Define whether land is designated for residential, commercial, mixed-use, or industrial purposes.
- **Land Use Tables:** Specify what types of structures or businesses are permitted in each zoning classification.
- **Building Setbacks & Height Restrictions:** Dictate how close a structure can be built to property lines and its maximum allowable height.
- **Density & Floor Area Ratio (FAR):** Determines how many units or how much building square footage can be developed.
- **Parking & Open Space Requirements:** Specify the minimum number of parking spaces and open areas required for different developments.
- **Overlay Districts:** Special zoning districts that impose

additional requirements, such as historic preservation, environmental protection, or urban design standards.

- **Zoning Ordinances:** Find the property's zoning classification and cross-check it with permitted land uses.
- **Land Use Tables:** Identify whether rezoning is necessary or if additional variances are required.
- **Municipal Codes for Building Requirements:** Research setbacks, max lot coverage, and development constraints.
- **Board of Zoning Adjustments (BZA):** If an exception is needed, check what approvals are required.
- **Rezoning & Variance Procedures:** Determine if a rezoning request is viable, how long it takes, and the likelihood of success.

Once you have a solid understanding of municipal codes, zoning regulations, and land use restrictions, the next critical step is conducting thorough due diligence on the property itself. Beyond reviewing maps and regulations, physical site conditions must be evaluated through various tests and documentation. This ensures that the land is suitable for development and uncovers any potential challenges that could delay or derail a project. Factors like soil stability, environmental hazards, flood risk, utility capacity, and title restrictions all play a crucial role in determining feasibility. Below is a list of essential due diligence documents and tests to consider before finalizing a land acquisition.

Survey with Topographic (Topo) Map

- A survey is a professionally measured map of the property showing its boundaries, physical features, and improvements.
- A topo survey includes elevation changes, slopes, and contour lines.
- Why it matters:
  - Confirms property boundaries to prevent disputes.

- o Identifies hills, low-lying areas, and natural drainage patterns that can impact grading and site layout.
- o Helps civil engineers design infrastructure like roads, stormwater systems, and foundation work.

Additional Tip:

- Always request a "Title Survey" to check for encroachments, setbacks, and easements, and ALWAYS review and compart the legal description.
- In steep-slope areas, ensure the survey includes geotechnical soil analysis (see below).

Soil Test (Geotechnical Analysis)

- Determines the composition, stability, and load-bearing capacity of the soil.
- Key soil factors analyzed:
  - o **Clay Content:** Can cause expansion and foundation shifting.
  - o **Rock Depth:** Determines excavation costs.
  - o **Permeability:** Affects septic system feasibility.
  - o **Contaminants:** Checks for hazardous materials.
- Why it matters:
  - o Prevents costly foundation repairs or structural failures.
  - o Determines whether soil improvement methods (e.g., compaction, pilings) are needed.

Additional Tip:

- In areas with unstable soil, additional testing, like boring samples, may be required.

Phase 1 Environmental Site Assessment (ESA)

- A preliminary environmental study that evaluates potential contamination risks.
- What it looks for:
  - Underground storage tanks (old gas stations, industrial sites).
  - Historical land use (factories, landfills, hazardous material storage).
  - Proximity to Superfund sites or other high-risk areas.
- Why it matters:
  - If contamination is found, a Phase 2 ESA (detailed testing) is required.
  - Helps avoid liability for costly environmental cleanups.

Phase 2 Environmental Site Assessment (ESA)

- Involves soil, groundwater, and air sampling to confirm contamination risks.
- Required if Phase 1 ESA identifies potential environmental hazards.
- Why it matters:
  - Can halt or delay development if contamination is found.
  - Remediation can cost hundreds of thousands if soil cleanup is required.

Additional Tip:

- If purchasing a site with past industrial or gas-station use, budget for potential remediation costs.

Traffic Study

- Evaluates road capacity, congestion, and required infrastructure improvements.

- What it includes:
  - **Traffic Counts:** Determines peak-hour congestion.
  - **Access Analysis:** Checks if turn lanes, signals, or road expansions are required.
  - **Projected Impact:** Determines how new development affects surrounding roads and whether traffic control and calming devices may be needed.
- Why it matters:
  - Helps avoid costly traffic impact fees.
  - Determines whether the site can handle the proposed number of homes or businesses.

Additional Tip:

- Some cities require developer-funded road improvements if traffic impact is significant.

Stream & Wetland Determination

- Identifies streams, wetlands, and floodplains that impact development.
- What it checks:
  - Blue-line streams (protected water bodies).
  - Flood zones & flood insurance requirements.
  - Wetlands mitigation requirements.
- Why it matters:
  - Developing on protected wetlands requires federal and state permits.
  - Can reduce buildable land area and increase stormwater management costs.
  - Determining between streams and wet weather conveyances.

Additional Tip:

- Check GIS & FEMA flood maps before making an offer.

Site Plan & Concept Plan

- A preliminary design showing how the land will be divided and developed, including roads, stormwater mitigation, and other requirements.
- Key elements:
  - Lot sizes and dimensions.
  - Road layouts.
  - Green space, sidewalks, and stormwater features.
- Why it matters:
  - Helps with zoning approvals and city planning discussions.
  - Provides realistic density estimates.

Additional Tip:

- Work with a civil engineer to get a sketch or character concept to bring to community meetings to help with your approvals.

Perc Test (Percolation Test)

- Determines septic system feasibility by measuring how quickly soil absorbs water.
- Why it matters:
  - Required for properties without municipal sewer access.
  - Some soils fail percolation tests, making land unsuitable for septic systems.

Infrastructure Cost Estimates (Hard & Soft Costs)

- **Hard Costs:** Physical construction (roads, sewer, utilities).

- **Soft Costs:** Legal fees, permitting, architecture, engineering.
- Why it matters:
  - Determines total development cost before purchasing land.
  - Helps investors estimate lot sales price & profit margins.

Title Work & Easement Review

- Confirms ownership history, liens, and encumbrances.
- Checks for:
  - Easements (access restrictions, utilities).
  - Encroachments (neighbors using part of the land).
  - Deed restrictions (zoning limitations, land use rules).
- Why it matters:
  - Unrecorded easements can reduce usable land area.
  - Liens can delay closing.

Water & Sewer Capacity Study

- Confirms adequate water & sewer access.
- Why it matters:
  - If no existing sewer capacity, developers may need to pay for expensive upgrades.
  - Alternative solutions: Private sewer systems or tapping into adjacent property utilities.

Additional Tip:

- Always request water/sewer maps from the local municipality.

Hydrotest (Flow Test)

- Measures water pressure & flow rate.
- Why it matters:
  - Required for multifamily and commercial developments.

- o If pressure is too low, booster pumps or larger pipes may be needed.

Stormwater Management Plan

- Determines how the property handles rainwater drainage.
- Why it matters:
  - o Prevents flooding, erosion, and environmental violations.
  - o Some cities require stormwater detention ponds or underground storage.

Not every due diligence item on this list will be necessary for every deal. The key is to assess what's relevant to your specific project before spending money on reports and studies that may not be needed. For example, a Phase 1 Environmental Site Assessment is critical for former industrial properties but may be unnecessary for a residential infill lot in a well-developed neighborhood. Similarly, a full traffic study might not be required for a small subdivision but would be essential for a large multifamily project. Your civil engineer will be one of your best guides in determining what's essential and what can be skipped. They'll typically have experience with similar projects in the area and can help you prioritize the most crucial studies based on zoning, site conditions, and city requirements. Always consult with your engineer and local permitting officials before ordering expensive reports, as they can often provide insight on what will be required for approvals and what may just be a "nice-to-have." Being strategic with due diligence will help you save money, avoid unnecessary delays, and keep your deal on track.

Understanding market research and due diligence is only the beginning of a successful land development strategy. Once you've identified a property, assessed its highest and best use, and determined viable exit strategies, the next step is navigating the **land entitlement and approval process**. This phase is where raw land is transformed into a legally buildable and marketable asset through zoning changes, site plan approvals, permits, and infrastructure planning. Entitlement is often the most critical and complex part of the development process, as it involves working with municipalities, planning commissions, engineers, and sometimes even local communities to secure the necessary approvals. In the next chapter, we will break down the entitlement process, from zoning and variances to subdivision approvals and utility planning, providing a roadmap to move from concept to reality.

CHAPTER 4
LAND ENTITLEMENT &
THE APPROVAL PROCESS

L and entitlement is one of the most crucial and complex aspects of real estate development. It involves obtaining the legal right to develop land for a specific use, whether that's residential, commercial, mixed-use, or something else entirely. The entitlement process is where concepts are transformed into buildable projects, requiring approvals from a variety of municipal agencies and stakeholders. This chapter will provide a detailed overview of the entitlement process, highlighting the importance of understanding local regulations, building relationships with city officials, and navigating the often complicated world of zoning, permits, and community input.

The land entitlement process is more than just following legal guidelines; it's about successfully collaborating with city planners, zoning officials, and municipal agencies that are essential to securing approvals. It requires developers to understand the inner workings of city departments like zoning, water, sewer, fire, and planning, as well as establishing strong relationships with decision-makers in these departments. A proactive approach to these relationships can make a significant difference when seeking approvals or negotiating project details.

THE IMPORTANCE OF RELATIONSHIPS WITH CITY OFFICIALS

Building positive relationships with city officials and departments is essential when working on land entitlement projects. Officials in zoning, water, fire, planning, and public works departments play critical roles in the approval process, as their support or objections can significantly impact your project's success. Developers who establish strong working relationships with these key players are more likely to receive timely feedback and guidance, which can help them avoid costly mistakes and potential roadblocks.

Equally important is understanding how the local council district operates in rezoning cases. Just because a project complies with zoning laws doesn't mean it will automatically gain council support. In many municipalities, city council members have the authority to vote on rezoning bills, and their support is often necessary to secure approval. It's vital to communicate the benefits of the project to both council members and the community, addressing potential concerns and emphasizing how the project aligns with the community's best interests.

It's also important to remember that the city officials and municipal employees you're working with are **W-2 employees**—they don't have a personal stake in your success, and they certainly aren't staying up at night worrying about your project. Their job is to enforce regulations, not to make things easier for you. That's why it's crucial to build goodwill. Bring them coffee, treat them with respect, and make them feel like they're part of something bigger. A little appreciation goes a long way—when they feel valued and included, they're more likely to take an interest in your project and help you navigate the system. Give them credit when things go smoothly, acknowledge their efforts, and they'll remember you the next time you need something approved.

BASIC ENTITLEMENT PROCESS

1. **Identify Zoning & Land Use:** The first step is to determine the current zoning of the property and its permitted land uses. This involves consulting local zoning ordinances and land use maps to understand what types of development are allowed.

2. **Pre-Application Meeting:** Many municipalities offer pre-application meetings with city planners to discuss the proposed project. This meeting provides developers with valuable feedback and identifies potential challenges early in the process. Most of the time, you will have to provide site plans and character concepts for what you are proposing.

3. **Rezoning or Variance Applications:** If the proposed use is not permitted under the current zoning, developers must apply for rezoning or a variance. This process often involves submitting detailed plans, including site layouts, architectural renderings, and descriptions of the project's impact on the surrounding area.

4. **Community Engagement:** Engaging with the local community and addressing their concerns is crucial. This often involves hosting community meetings, answering questions, and demonstrating how the project will benefit the neighborhood.

5. **Navigating Public Hearings:** Rezoning applications often require public hearings before planning commissions, zoning boards, or city councils. Developers must present their case, respond to objections, and make a compelling argument for why the project should be approved.

6. **Board of Zoning Appeals (BZA) Process:** If a variance or exception is needed, the project may go before the BZA. This board evaluates requests for deviations from zoning regulations, such as building height or setback requirements. Whenever you have to present to the BZA, it's essential to

show some type of hardship against the property, whether it be setbacks or other constrictions.

7. **Securing Site Plan Approvals:** Developers must submit detailed site plans for approval, which typically include layouts for utilities, roads, stormwater management, and landscaping. These plans must meet municipal standards and may require revisions based on feedback from city planners.

8. **Obtaining Special Permits:** Some projects require special permits or conditional use permits. These permits allow certain types of development in specific zones, provided the project meets additional criteria set by the municipality.

LEGAL & REGULATORY HURDLES TO WATCH OUT FOR

- **Easements & Encroachments:** Understanding existing easements and potential encroachments is crucial, as they can impact site design and development plans.
- **Bylaws & HOA Restrictions:** Some properties may be subject to bylaws or homeowner association (HOA) restrictions that limit what can be built on the land.
- **Overlay Districts & PUDs:** Overlay districts and PUDs have specific design and development requirements that must be met. These often involve additional layers of review and approval.

This chapter may be short, but it is extremely important because every area is different and these basic understandings should give you an outline to look out for. Also keep in mind the entitlement process isn't always **linear**—every municipality has its own **rules, procedures,** and **timelines**, and approvals can vary significantly depending on location. What works in one city may not apply in another, which is why it's crucial to **schedule a meeting with the city planner or key officials** early on to understand the specific steps for your project. This not only helps **streamline your timeline** but also builds rapport with decision-makers. Another useful practice is **finding the city's YouTube channel** (if they have one) to watch past planning and zoning hearings, city council meetings, and Board of Zoning Appeals (BZA) sessions. Studying how these meetings unfold, what **other developers do right or wrong**, and how different officials **respond to proposals** can help you prepare your strategy and avoid costly mistakes.

Additionally, take the time to **understand the personality types and decision-making styles** of those involved in the approval process. Some officials are **strict rule-followers**, while others may be more **flexible, with creative solutions**. Tailoring your approach accordingly can be the difference between an approval and a rejection. Last, don't overlook the **neighbors**—especially in residential areas, their support (or opposition) can play a major role in how smoothly your project moves forward. **Building relationships early** and addressing concerns upfront can prevent resistance later and even turn potential opponents into allies.

Speaking of allies, building the right team is one of the most crucial aspects of a successful land development project. Each professional plays a distinct role in ensuring the project moves forward efficiently while navigating the complexities of zoning, infrastructure, and municipal approvals. Relationships with city officials, engineers, and consultants are just as important as the technical expertise each team member brings.

Having a **strong team in place before acquisition** ensures that you can properly evaluate deals, mitigate risks, and streamline the entitlement process, ultimately leading to a more profitable and successful project.

TEAM MEMBERS NEEDED FOR SUCCESS

1. *Surveyor*

Role: Surveyors are responsible for determining the exact boundaries of the land, existing structures, easements, and topographical features. They conduct boundary surveys, topographic surveys, and ALTA/NSPS surveys, which are often required by lenders or title companies.

Why They're Important:

- Identify legal property boundaries and potential encroachments.
- Help with zoning, lot subdivision, and site planning.
- Required for title insurance and land transactions.

2. *Civil Engineer*

Role: Civil engineers assess and design the infrastructure needed for development, such as roads, drainage, sewer, and water systems. They work closely with surveyors and environmental consultants to ensure compliance with municipal codes.

Why They're Important:

- Prepare grading and drainage plans.
- Ensure infrastructure feasibility and regulatory compliance.
- Provide necessary calculations for stormwater management and utility planning.

3. *Land Use Attorney / Zoning Attorney*

Role: These professionals specialize in land use regulations, zoning laws, and real estate transactions. They help navigate municipal codes, rezoning applications, and land use disputes.

Why They're Important:

- Assist with rezoning, variances, and special permits.
- Help negotiate with city officials and community groups.
- Advise on easements, encroachments, and legal restrictions.

4. *Architect / Draftsman*

Role: Architects and draftsmen design conceptual site plans, building layouts, and architectural renderings. They ensure that the project aligns with zoning codes and design guidelines.

Why They're Important:

- Help with preliminary site plan approvals.
- Assist in maximizing land use efficiency.
- Ensure compliance with aesthetic and functional design regulations.

5. *Environmental Consultant*

Role: Environmental consultants assess site conditions for potential contamination, wetland impacts, and regulatory compliance.

Why They're Important:

- Conduct **Phase 1 & Phase 2 Environmental Assessments** to identify contamination risks.
- Determine wetland delineations and mitigations.
- Provide reports needed for municipal approvals.

6. *Real Estate Agent / Commercial Broker*

Role: Agents and brokers assist in acquiring land, marketing entitled projects, and identifying potential end users or investors.

Why They're Important:

- Help identify undervalued land opportunities.
- Facilitate land transactions and negotiations.
- Provide market insights and exit strategies.

7. *General Contractor & Subcontractors*

Role: General contractors oversee the construction phase, while subcontractors handle specific aspects such as grading, plumbing, electrical, and paving.

Why They're Important:

- Manage cost estimates, schedules, and permitting.
- Coordinate with engineers and city inspectors.
- Ensure proper execution of construction plans.

8. *Title Company*

Role: Title companies ensure that the land title is clear of liens, encroachments, or other legal restrictions that could impact development.

Why They're Important:

- Provide title insurance and legal documentation.
- Handle escrow and closing for transactions.
- Identify potential title defects before acquisition.

9. *Municipal Officials & Planning Department*

Role: City and county officials, including zoning administrators, planning commission members, and council members, oversee the entitlement and permitting process.

Why They're Important:

- Determine zoning and land use approvals.
- Control permit issuance and development compliance.
- Play a key role in rezoning and variance approvals.

10. *Traffic Engineer*

Role: Traffic engineers analyze the impact of new developments on local road networks and ensure compliance with transportation regulations.

Why They're Important:

- Conduct **traffic impact studies** required for permitting.
- Help determine road access points and necessary improvements.
- Assist in mitigating congestion concerns from city planners.

11. *Utility Companies & Engineers*

Role: These professionals assess water, sewer, electric, and gas availability and design infrastructure for new developments.

Why They're Important:

- Ensure adequate utility capacity.
- Help in coordinating service connections and easements.
- Avoid costly infrastructure miscalculations.

12. *Public Relations / Community Liaison*

Role: These team members help engage with the local community, address concerns, and gain public support for rezoning or entitlement applications.

Why They're Important:

- Help mitigate opposition at public hearings.
- Build relationships with local officials and stakeholders.
- Increase project approval chances through strategic communication.

13. *Financial & Funding Partners*

Role: Lenders, private investors, and institutional capital sources provide the necessary financing to acquire and develop land.

Why They're Important:

- Secure acquisition and construction loans.
- Provide capital for infrastructure improvements.
- Help structure deals with creative financing solutions.

14. *Land Planner / Urban Designer*

Role: Land planners help design the layout of subdivisions, mixed-use developments, and master-planned communities.

Why They're Important:

- Maximize land use efficiency.
- Help navigate overlay districts and community master plans.
- Work closely with municipalities on future land use goals.

15. *Soil Scientist / Geotechnical Engineer*

Role: These experts evaluate soil conditions for structural stability and drainage requirements.

Why They're Important:

- Conduct **soil tests** to determine load-bearing capacity.
- Identify potential contamination or hazardous materials.
- Provide recommendations for foundation and site preparation.

16. *Floodplain Manager & Environmental Permitting*

Role: This role involves evaluating flood risks and securing permits related to wetlands, stormwater, and environmental conservation.

Why They're Important:

- Ensure compliance with FEMA floodplain regulations.
- Identify necessary mitigations for stormwater drainage.
- Secure permits for wetland disturbances.

17. *HOA Consultant & Property Management*

Role: If the development includes a **Homeowners Association (HOA),** consultants can help structure the HOA and long-term maintenance plans.

Why They're Important:

- Help establish HOA bylaws and CC&Rs.
- Ensure proper budgeting for community maintenance.
- Provide strategies for long-term sustainability.

18. *Real Estate Attorney*

Role: Real estate attorneys negotiate and draft contracts, ensure clean title transfers, and help structure the purchase and sale agreements.

Why They're Important: They protect you from legal risks, ensure contracts are properly structured, and negotiate terms that favor your interests.

Before you start reaching out and hiring professionals, make sure you actually have a deal in place. Most industry experts won't provide valuable insights without a formal engagement, and you don't want to waste their time—or yours—by jumping the gun. When it is time to bring them on board, clarify whether they charge by billable hours or a project-based fee, as this can significantly impact your budget. Be upfront about expected timelines, and whenever possible, work with professionals who have experience in the specific city or county where your project is located. While hiring someone from out of town might seem cost-effective, it often leads to delays and missteps since they lack established relationships with key city officials and may not fully understand the local approval process. The best vendors have spent years navigating municipal regulations, building rapport with decision-makers, and learning the nuances of the local system—this experience can be the difference between a smooth project and a bureaucratic nightmare.

Assembling the right team is essential for successfully navigating the land entitlement and development process, but timing is just as critical. Having knowledgeable professionals in place will streamline approvals, prevent costly mistakes, and help avoid unnecessary delays. However, a strategic approach is key—securing a deal first ensures you're making efficient use of time and resources. Once you've built the right team and understand their roles, the next step is structuring and negotiating contracts. This is where all the groundwork you've laid starts to take shape, and well-crafted agreements are crucial to keeping deals on track while protecting your interests. Now, let's dive into the negotiation process and explore how to structure contracts that work to your advantage.

CHAPTER 5

CONTRACTS AND NEGOTIATION

WHY WE SAVED CONTRACT NEGOTIATION FOR LAST

We've deliberately placed contract negotiation later in the book because every step we've taken so far—market research, due diligence, entitlement, team-building, and approvals—determines how we negotiate the deal. Without understanding these elements, you'll be making offers blindly, risking your capital, and potentially tying yourself into contracts that are unfeasible, unprofitable, or downright disastrous. Now that we've covered the necessary groundwork, we can confidently approach contract negotiations, using this knowledge to craft agreements that protect our investment while ensuring the deal remains attractive and executable.

Let's break down the basics of what we've learned and how it all comes together when it's time to structure a deal.

UNDERSTANDING MARKET RESEARCH: THE FOUNDATION OF PRICING AND DEMAND

Before making an offer on any piece of land, we have to **analyze the market** and determine what the land is actually worth—not just today, but years from now when the final product is built and sold. This involves studying:

- **Proposed Units in the Pipeline:** How much competition will we face once the project is ready?
- **Current Market Demand and Price Trends:** Are homes or commercial buildings selling fast? At what price per square foot?
- **Who the Major Developers Are in the Area:** Are national home builders buying? What product are they building?
- **Development Trackers and Permitting Records:** What's already approved in the area, and how might that impact our project?
- **Sewer Capacity and Infrastructure Limitations:** Can the area support additional units, or will we need costly upgrades?
- **What Builders Are Paying for Land:** We must ensure that our land cost aligns with market comparables—raw land prices before infrastructure vs. finished pad prices after improvements.

Market research tells us **what the land's highest and best use is, who our buyers will be,** and **what they'll be willing to pay for a developed lot.** This allows us to negotiate from a position of **knowledge, not speculation.**

DUE DILIGENCE: IDENTIFYING RISKS BEFORE MAKING AN OFFER

Once we identify a potential deal, we **don't just throw out a number** —we conduct due diligence first. We've learned that due diligence means:

- **Checking Zoning Regulations:** Can we build what we need to make the deal profitable, or do we need a rezone?
- **Reviewing Comprehensive Plans and Overlays:** What's the

city's long-term vision for the area? Will they support our project?

- **Using GIS Mapping Tools:** Is the site on a floodplain, conservation area, or steep slope? Will these affect feasibility?
- **Evaluating Utilities and Infrastructure:** Does the site have sewer, water, and stormwater access, or will we have to install expensive upgrades?
- **Reviewing Title Work:** Are there easements, encroachments, or other legal issues that could impact the project?
- **Performing Environmental Assessments:** Is the site contaminated, requiring remediation before development?

This step protects us from **signing a contract without knowing exactly what we are getting into**. If we find major issues, we can either **renegotiate the price, adjust contract terms**, or **walk away altogether.**

Another thing to remember is if your land is vacant, it's crucial to **stay proactive in maintaining and protecting it** to avoid unnecessary delays and potential legal issues. **Regularly cutting the grass, clearing debris, and keeping the property in compliance with local ordinances** will help you **avoid code violations** that could lead to costly fines or municipal liens. Additionally, make sure **all property taxes are paid on time**—delinquent taxes can quickly turn into a bigger problem, sometimes even leading to foreclosure by the city or county. It's also **wise to invest in a general liability insurance policy** to protect yourself from potential lawsuits due to injuries or accidents on the property, such as a trespasser getting hurt or an unauthorized structure collapsing.

If squatters take up residence on the land, **do not attempt to remove them yourself**—this can lead to legal repercussions. Instead, follow the **proper legal eviction process based on local laws**, as some areas have strong tenant protections, even for illegal occupants. Failing to go through the correct legal channels could **cause unnecessary holdups** when it's time to sell or develop the property. Additionally, if you receive any **notices of code violations**, address them immediately to prevent further complications that could affect entitlements, financing, or resale value. By staying on top of these basic responsibilities, you ensure that your land remains a **clean, compliant, and valuable asset** without any unexpected roadblocks. It's good to make friends with the code enforcement department!

ENTITLEMENT AND THE APPROVAL PROCESS: UNDERSTANDING TIMELINES AND POLITICAL RISKS

Many new investors assume that just because land is zoned a certain way, they can automatically build. But as we covered in the entitlement process, **zoning and approvals are fluid and political**.

- **Rezoning doesn't have to be supported just because it makes sense.** We saw how local council members can support or reject projects based on their own agendas, community sentiment, or other political factors.
- **Every municipality has different rules.** Some require three public hearings, others require community meetings, and some have unwritten policies that can stall or speed up approvals.
- **The Board of Zoning Appeals (BZA) process is complex.** Variances for things like setbacks, parking, and overlays require strategic planning and often require legal help.
- **Knowing the right people makes a difference.** Relationships with zoning officials, planning staff, and city council members are critical. We learned that **these are W2 employees who don't care about our financial success**—but treating them well, giving them credit, and making them feel involved **can go a long way in getting approvals through smoothly**.
- **Timelines can make or break a deal.** We discussed how zoning changes can take anywhere from six months to two years and how understanding city processes in advance helps us build realistic contract contingencies.

This section taught us that **contract timelines must reflect the reality of entitlements and approvals**. If we don't factor these in, we could be stuck with land that's not ready for development or forced to close before we have the necessary approvals.

ASSEMBLING THE RIGHT TEAM

Even the best-negotiated contract means nothing if we don't have the right people in place to get the deal done. Throughout this book, we identified the **key players needed for a successful project**, including:

- **Attorneys:** To structure contracts, negotiate terms, and ensure we're legally protected.
- **Civil Engineers:** To assess feasibility, create site plans, and help navigate municipal approvals.
- **Land Use Attorneys:** To handle rezoning, variances, and legal entitlements.
- **Surveyors:** To conduct ALTA surveys and boundary surveys, and ensure legal descriptions are accurate.
- **Environmental Consultants:** To assess soil conditions, wetlands, and any required remediation.
- **Real Estate Agents and Brokers:** To find off-market deals, connect us with buyers, and help evaluate pricing.
- **Title Companies:** To ensure clean title, clear encumbrances, and manage closings.

One of the biggest mistakes a new investor can make is **hiring these professionals too soon**. We discussed how most won't provide real value until there's a contract in place and how we should always clarify whether they're billing by the hour or per project.

STRUCTURING THE RIGHT EXIT STRATEGY

The last critical piece before finalizing a contract is confirming our exit strategy—or, hopefully, strategies. I know I said that it was important to have your exit strategy in place before beginning anything, but it's even more important to validate it before going into contract negotiations. We've reviewed multiple exit strategies to ensure we have a clear and viable path forward, mitigating risk and maximizing our potential returns before committing to the deal. We reviewed multiple exit strategies, including:

- **Wholesaling the Land:** Assigning the contract for a fee.
- **JV with a Builder or Investor:** Partnering to bring the project to completion.
- **Holding for Appreciation:** Land banking for long-term gains.
- **Selling Entitled Land:** Getting approvals and flipping the paper.
- **Developing and Building:** Taking the project vertical ourselves.
- **Subdividing and Selling in Pieces:** Maximizing returns by breaking up the land.

Every deal must have at least **one solid exit strategy—preferably two or three**. This ensures we don't get stuck if market conditions change or if a buyer backs out.

BRINGING IT ALL TOGETHER: NEGOTIATION

Now that we've laid the groundwork, we can **negotiate contracts with confidence**. We understand:

✅ **The Market Value of the Land:** So we don't overpay.

✅ **Potential Risks and Challenges:** So we negotiate realistic contingencies.

✅ **Zoning and Entitlements:** So we structure contracts with proper timelines.

✅ **The Key Players Needed to Close the Deal:** So we can execute efficiently.

☑ **Exit Strategies:** So we always have a plan for getting out profitably.

With this knowledge, we're **not just throwing out numbers and hoping for the best**. We're crafting smart, strategic contracts that give us control while protecting us from unnecessary risk.

Now, let's dive into **the art of negotiating contracts**.

Because of these timeframes, your contract **must reflect realistic expectations** for due diligence, rezoning contingencies, and the ultimate closing date. **If approvals take longer than expected and your contract doesn't provide enough time, you could be forced to close on an undevelopable property or risk losing your earnest money.** This is why **negotiating extensions or having clear contingency clauses is critical** when structuring your deal.

DETERMINING CONTRACT TIMELINES BASED ON APPROVAL PROCESSES AND TIMELINES

Before structuring any land contract, **the first and most important factor to consider is the approval timeline**. The length of time required to secure necessary entitlements, rezoning, or permits will directly impact **how long you need before closing** and what contingency periods should be included in your contract.

Rezoning and entitlement timelines **vary significantly based on the municipality, the complexity of the request**, and **local government schedules**. A straightforward rezoning might take **as little as four months**, while a more complex case requiring community engagement, environmental studies, or infrastructure planning can extend well beyond **a year or more**. It's essential to refer to the planning department and craft your contract contingencies based upon those timelines. For instance, if the planning department says six months, you should make your contingency six months plus built-in extensions, just in case there is a delay, and many times, there is. One thing to also keep in mind is the time it takes to be "signed into legislation." If it takes a week or more to get the bill signed, you need to make sure that language is in your contingency. The reason is that banks won't lend, and it's not done until it's done when that bill gets signed. You can also build in earnest money or hard money (when it's no longer refundable) milestones so you can help keep the seller engaged throughout the process if they are pushing back on their timelines. Show the seller the charts or emails that show them your timeline is legit. Make sure you are transparent with the seller. Be upfront about any physical approvals you may need: for example, easements or sewers or the city putting in certain utilities. When you have a hearing, make sure you let them know the result. Everything will go on pubic record, so it's best they hear the results from you instead of seeing it online or in the paper. You need to keep the seller trusting you throughout the process; it is very important.

When working with an assignment or an end buyer contract, the same principles apply as when structuring your initial agreement with the seller—but with even more layers of complexity. You must determine what the end buyer requires in terms of **timelines**, **contingencies**, **financing**, and **approvals** while also aligning with the city's entitlement and approval process. This is critical because if your end buyer's needs don't match up with your original seller contract, you could find yourself squeezed between two conflicting agreements, leading to unnecessary risk and financial exposure.

One of the **biggest mistakes** investors make when structuring an assignment or end buyer contract is failing to account for financing contingencies. Many **end buyers may initially sign a contract as a "cash buyer"** but later decide to secure financing. If this happens and they use a bank loan, **you must know the bank's lending requirements**, as they can dictate additional due diligence, title clearance, and legal reviews, which can **extend timelines significantly**. Banks often require new title work and underwriting, and many do **not** lend on assignments. If this happens, **you may be forced to draft a whole new contract**, introducing new legal complexities and potential delays.

To mitigate these risks, **make sure your contract with the seller has enough contingencies and built-in extensions** to allow flexibility in case the end buyer's funding process takes longer than expected. Also, **ensure that there is specific language in both your assignment and end buyer PSA (Purchase and Sale Agreement)** to protect you from losing your earnest money or damaging your credibility with the seller. In cases where financing delays are common, **consider structuring your exit as a double closing** rather than a simple assignment. While this requires additional transactional funding or short-term bridge financing, it ensures the deal remains intact and that you maintain control over the closing process.

Additionally, **communication is key**. Keep the seller informed about the process, and if there are necessary city approvals, easements, or infrastructure changes required, make sure both the seller and the end buyer are aware. If there are delays at city hearings or approvals take longer than expected, **let the seller know immediately** rather than allowing them to find out from public records. Keeping the seller engaged and transparent throughout the process **builds trust and increases the likelihood of cooperation** if additional time or adjustments are needed.

Ultimately, the goal is to **structure the deal in a way that allows you to remain in control while minimizing your risk**. Anticipate potential financing and approval roadblocks, structure contingencies accordingly, and work with attorneys experienced in assignments and double closings to **ensure that every contract aligns properly** with the seller's and end buyer's expectations.

General Breakdown on Timelines for Entitlements

- **Pre-Application Meeting:** 1–2 weeks to schedule. Many municipalities require this step to discuss feasibility with planning staff before submitting an application.
- **Rezoning Application Submission:** 2–4 weeks for initial

review. The city reviews the application for completeness before moving it to the next stage.

- **Public Hearing & Community Meetings:** 2–3 months. Community feedback can influence whether your project is supported or opposed. This is a critical period that may require adjustments to your proposal.
- **City Council or Zoning Board Vote:** 1–3 months after public hearings. The final decision on your rezoning request happens in a series of votes, with some requiring multiple readings before approval.
- **Final Approvals & Permits:** 1–6 months, depending on the complexity of the project. This phase includes site plan approval, building permits, and any required environmental or infrastructure sign-offs.

UNDERSTANDING OF THE CONTRACT PARTIES

1. *The Seller & Buyer (You)*

This is the **initial and most critical negotiation**, as it sets the foundation for the entire transaction.

- You, as the buyer, will typically **present an LOI (Letter of Intent)** to outline basic terms such as purchase price, due diligence periods, contingencies, and closing timeline.
- Once terms are agreed upon, a **formal PSA** is drafted— usually by **your real estate attorney**.
- The **seller may have their own attorney** review the PSA, which could **extend the negotiation period** while they counter terms or request modifications.
- If the property requires rezoning or entitlements, **your contract timelines should align with municipal approval processes** to avoid being stuck in a deal without the necessary approvals.

The goal is to **structure the contract with enough contingencies** to protect your earnest money and give you flexibility in case delays arise.

2. *The Buyer (You) & Assignor (If Assigning the Contract)*

If you intend to assign the contract rather than close on the property yourself, you're now **negotiating with an end buyer (the assignor)**.

- The **assignor (your end buyer)** may submit their own LOI or simply accept the assignment terms you propose.
- In most cases, the **PSA must allow for assignment**, or you may need to negotiate an **Assignment of Contract Agreement** separately.
- If the end buyer intends to **use financing**, this can complicate the assignment process, as **many banks won't lend on assignments**. In such cases, a new contract may need to be drafted instead of a simple assignment.

You must **ensure that the end buyer's timeline aligns with your PSA with the original seller**—otherwise, you could be **stuck in a contract without a buyer** or risk default.

3. *The Buyer (You) & End Buyer (The Person Buying from You)*

When you have a property under contract and are **reselling it to an end buyer**, whether through an **assignment** or **double closing**, additional factors come into play.

- **Corporate Buyers & Home Builders:** Large corporations, **national home builders**, or **syndicators** typically **have their own legal teams** to review contracts. Their attorneys will likely go back and forth with your attorney to **negotiate terms, assignment clauses**, or **financing contingencies**.

- **JV & Syndicators:** If the end buyer is **raising capital** through a **joint venture (JV) or syndication**, expect delays in contract execution, as **their investors or financial partners must approve the deal** before moving forward.
- **Bank Requirements:** If the end buyer is **securing financing**, you must determine **whether their bank will require a fresh contract** rather than a simple assignment. Many lenders require **a new purchase agreement and their own title search** before funding.
- **Title & Legal Review:** Attorneys from both sides will review title, contract terms, and zoning approvals to ensure a **clear and marketable title** before closing.

Since multiple attorneys may be involved in **reviewing documents**, **conducting title searches**, and **finalizing terms**, the process can take **several weeks or even months**. You need to **factor this into your contract timelines** and ensure that your seller's contract allows for enough time to get through these negotiations without defaulting. Putting due diligence start dates from these milestones could help mitigate that risk.

UNDERSTANDING KEY CONTRACT TYPES FOR LAND ACQUISITIONS

When structuring a land deal, having the right contract type in place is essential to protecting both your interests and your timeline. There are two main approaches to securing a property: a **General PSA** and a **LOI leading to a formal PSA**.

1. *General PSA*

This is the standard contract used when purchasing a property directly from a seller. It includes **the agreed-upon purchase price, contingencies, due diligence periods,** and **closing timelines**. Since every deal is different, you may need to **modify the contract language to fit specific terms you and the seller negotiate**.

A good practice is to have an attorney draft a **template PSA** that you can keep on hand and use whenever needed. I prefer to have **two separate PSA versions—one as a buyer and one as a seller**. The key difference lies in the **default section**:

- **Buyer PSA:** If the seller fails to close after all contingencies are met, I include a clause that allows me to **sue for performance**. This holds the seller accountable and ensures they cannot back out after an agreement is reached.
- **Seller PSA:** If I am selling the land, I include language stating that the **earnest money satisfies and holds harmless** so that the buyer cannot sue me if the deal falls apart. This protects against drawn-out legal disputes.

Since a **PSA is legally binding once signed**, it's important to make sure it aligns with your deal structure before executing. **Your attorney can draft this once, and you can reuse it across multiple deals with minor modifications.**

2. *LOI Leading to a PSA*

An **LOI is a non-binding agreement** that outlines the basic deal terms before drafting a full contract. This method is useful in complex transactions or when working with institutional sellers, developers, or corporate landowners.

The LOI will include key terms such as:

- Purchase price
- Due diligence period
- Closing timeline
- Deposits
- Any special conditions (entitlement approvals, financing contingencies, etc.)

Once the LOI is agreed upon, a formal **PSA is drafted by the attorney** and sent to the seller's attorney for review. If the seller does not have an attorney, just send it directly to them.

One thing to keep in mind is that **negotiating an LOI can take a few weeks, and drafting a PSA can take even longer, depending on the complexity of the deal**. Some PSAs—especially for larger or entitlement-heavy deals—can take weeks or even months of back-and-forth negotiations before they are finalized.

Choosing between a direct PSA or an LOI first depends on the deal and the seller. If you're dealing with a **motivated seller who is ready to move quickly, a PSA may be the fastest route**. However, if the seller needs time to review terms or if there are multiple decision-makers involved (such as corporate sellers, family members, or partnerships), starting with an LOI may be the better approach.

NAVIGATING CONTRACTS WITH THE END BUYER

Once you have secured a property under contract with the seller, the next step is structuring an agreement with the **end buyer**. The complexity of this process depends on **whether you have already closed on the property** or if you are **still under contract with the seller**. Each scenario requires a different approach, and failing to structure the right contract **can lead to delays, financing issues**, or even **legal conflicts**.

1. *Direct Assignment (Selling a Contract Without Closing)*

If you have **not yet closed on the property**, you are essentially selling your **contract rights** to the end buyer, not the actual property. This is commonly done through an **assignment contract**, which transfers your interest in the original PSA to the buyer for a fee.

However, **not all end buyers can use assigned contracts**. If the buyer is using bank financing, you must ensure that their lender allows assignments. Many traditional banks and institutional lenders **will not fund deals structured as assignments** because they require the end buyer to have full ownership and run their own title searches before closing.

This means if you planned to assign the deal but the buyer's bank will not allow it, **you now have to completely restructure the contract**. Instead of a simple assignment, the buyer will need to execute a **brand-new PSA** with you as the seller, effectively replacing your contract with the original seller.

When assignments and end buyers enter the equation, the transaction can quickly become more complex, requiring careful navigation to protect your interests and ensure a smooth closing. One of the first challenges is that the buyer's attorney will likely get involved to negotiate new terms, which can lead to additional back-and-forth and potential modifications to the original agreement. Your attorney plays a crucial role in safeguarding your position and making sure you are paid the agreed-upon fee despite any changes that may arise. Lender requirements also add another layer of complexity—many banks will require updated title work, appraisals, or additional financing contingencies, which can delay closing. If the end buyer is using financing instead of cash, it's important to understand the bank's requirements in advance to avoid last-minute surprises. Additionally, if your end buyer is from a different state, legal conflicts can arise between local laws and the governing contract laws of their home state. This requires careful legal review to determine which state's laws apply and how they may impact timelines, contract enforceability, and any necessary approvals. These moving parts can create unexpected challenges, which is why having the right legal team and a clear contract structure in place is essential when navigating assignments and working with end buyers. This is why **choosing the right attorney is critical**—one who understands how to navigate contract assignments, lender requirements, and legal variations across jurisdictions.

2. *Double Closing (When Assignments Are Not Possible)*

If the end buyer cannot purchase via an assignment, **you may need to do a double closing**. A double closing involves **you purchasing the property from the seller first** and then selling it to the end buyer immediately after. These will cause delays and additional fees to the transaction.

The challenge here is that you need **funding to close the first transaction**. If you don't have the capital readily available, you may need:

1. **Transactional Funding:** Short-term capital that covers the purchase price just long enough to close the second transaction.
2. **Bridge Financing:** A temporary loan that gives you time to secure longer-term financing or complete the resale to the end buyer.
3. **Investor Partnerships:** Bringing in outside capital partners who are willing to provide funds in exchange for a share of the profits. They can be private equity investors for interest of equity of a JV partner. You may have to negotiate what makes sense for your deal.

This method introduces additional risk because:

- You **must be certain your end buyer will close**, or you will be stuck owning the property.
- If financing delays occur, **you may need to hold the property longer than expected**. If you know you may own the property for a short period of time, it's good to prepare for a long-term hold.
- **Additional closing costs** are incurred because two separate transactions take place.

If a double-close is required, you should consider **charging the end buyer a premium** to cover the added complexity, risk, and financing costs. This should be negotiated upfront before you commit to closing the first contract.

WORKING WITH BUYERS USING FINANCING

Buyers who use bank financing introduce **a different set of challenges.** If they originally sign a **cash purchase agreement** but later decide to use financing, you must ensure:

1. **The lender's requirements are met** (title work, inspections, environmental reviews, etc.).
2. **The contract timelines align** with the lender's closing process, which can take weeks or months.
3. **The buyer can actually qualify for financing**—a common issue that causes last-minute deal collapses.

Most banks **will not lend on an assignment contract**, meaning a new contract must be drafted. **This triggers attorney reviews and lender underwriting, which can introduce delays or unforeseen requirements.**

Some key risks to watch for when dealing with bank-financed buyers:

- If the bank requires **ownership before running title work,** you may be forced to close first.
- If the bank needs to **verify infrastructure improvements or entitlements**, additional approvals may be needed before funding.
- The buyer's financing **may not be fully secured**, putting your sale at risk if they cannot close.

To prevent last-minute surprises, always confirm with the end buyer how they plan to fund the deal upfront. One thing to always do is have conversations with the end buyer's lender. It's not like in a residential real estate transaction where you get a pre-approval letter. Also, be sure the lender has closed similar land deals in the past.

THIS IS YOUR OPPORTUNITY!

There's no single way to succeed with land development, and I can't wait to hear about the approach you take. Please take a moment to leave your feedback online so I can hear about your plans and other potential investors can find this book more easily.

Simply by sharing your honest opinion of this book and your thoughts at this stage of your journey, you'll show other potential developers where they can find the guidance they're looking for—and you'll let me know how this has helped you at the same time.

LET'S HEAR WHAT YOU THINK!

Thank you so much for your support. I have every faith in your ability to secure generational wealth through land development, and I'm excited for the opportunities ahead of you.

Scan the QR code below to leave your review on Amazon.

CONCLUSION

Throughout this book, we've broken down the complex world of land acquisition and entitlement into actionable steps that anyone can follow. From identifying opportunities and conducting due diligence to navigating municipal approvals and structuring contracts, you now have a clear roadmap for approaching land deals strategically. While the process may seem intricate at first, the reality is that success in land development doesn't require a special degree or years of experience—it requires persistence, attention to detail, and the ability to put the right team in place. Every challenge in this business has a solution, and with the right knowledge and mindset, you can tackle any deal confidently.

One of the biggest takeaways is that real estate development is not about simply flipping land or pushing paper—it's about solving problems. Whether you're working through zoning changes, negotiating contract terms, or structuring creative exit strategies, every deal is an opportunity to learn and refine your skills. The best developers and investors aren't the ones who avoid problems; they're the ones who know how to overcome them. With each deal, you gain more confidence, build stronger relationships, and develop a sharper eye for opportunity.

Now, it's time to move from theory to practice. In these Appendices, we'll walk through real case studies—actual deals that have been successfully executed. These examples will give you a firsthand look at how different strategies play out in the real world, how obstacles were overcome, and what lessons were learned along the way. No two deals are the same, and while the process can sometimes seem unpredictable, the key is being adaptable and understanding how to adjust when needed.

If anything, these case studies will prove that land development isn't reserved for large corporations or industry veterans—it's something **you** can do. By breaking deals down into manageable steps and approaching them with the right information, **anyone can succeed in this space**. Let's dive in and see how these concepts apply in actual transactions.

Happy dirt investing!

CASE STUDIES: REAL DEALS, REAL LESSONS

The best way to truly understand the land development process is by looking at real deals—transactions that have gone through the ups and downs, the unexpected hurdles, and the ultimate payoffs. Every case study in this chapter is a deal I've personally been involved in, and these are not hypothetical scenarios or cherry-picked successes. These are real transactions that you can look up online through tax records, zoning approvals, and recorded sales. Some were massive wins, others had missed opportunities, and all of them provided invaluable lessons that shaped how I approach land entitlement and development today. By breaking down each deal, you'll get a front-row seat to the strategies, negotiations, challenges, and pivots that led to the final outcomes.

CASE STUDY: 503 W TRINITY, NASHVILLE, TN, 37207

Purchase Price: $2,000,000 | Exit Price: $3,450,000

This deal is a perfect example of how **not having to go all the way** in the entitlement process can lead to a **faster and more profitable exit**. Rather than fully entitling the project with a finalized site plan, we strategically pursued a **Preliminary Specific Plan (SP)**, allowing us to maximize zoning flexibility while minimizing time spent on approvals. A **Specific Plan (SP)** zoning designation can make a property much harder to sell because it requires every detail—zoning, land use, site layout, architecture, materials, and even landscaping—to be pre-negotiated and approved by the city. Unlike standard zoning, where a new owner can adjust their development within the given regulations, an SP locks in an exact plan. If a new buyer wants to make any changes, they must go through the entire SP amendment process, which often takes longer than an initial rezoning. This added layer of approval can deter buyers who need flexibility, making SP-zoned properties more challenging to market and sell.

The Strategy: Leveraging Preliminary Entitlements for a Faster Exit

We initially sourced this deal from a **savvy investor** who understood zoning intricacies and was open to working with us to get the property **rezoned and positioned for an optimal exit**.

The site was already **split-zoned**, meaning part of it had one zoning designation while the other portion had another. Instead of going through the lengthy and expensive process of rezoning everything, we **worked with the city to secure a Preliminary SP**.

Understanding Preliminary SP vs. Final SP

- **Preliminary SP:** Establishes the **general framework** of the project, including zoning allocations, road layouts, and density approvals. This allows a future buyer to move forward with confidence, knowing that the groundwork is set.
- **Final SP:** Includes detailed **site plans, architectural drawings, utility layouts, parking requirements**, and **environmental considerations**. This takes significantly more time and capital to secure.

By **stopping at the Preliminary SP stage**, we saved the next buyer **at least twelve to eighteen months** or **more** of entitlement work. This meant they could immediately move forward with either **vertical development** or selling portions of the land separately.

Breaking Down the Plan

Our entitlement process allocated the property into:

1. **60 residential units in the back** (maximizing the density in a way that fit the city's long-term vision)

2. **Box-unit commercial layouts in the front** (basic footprints that allowed for flexible commercial use)

By keeping the entitlement process **simple and open-ended**, we gave the future developer multiple options:

- Build the commercial upfront and generate early cash flow
- Sell off the commercial pad separately to **recoup capital quickly**
- Proceed directly with multifamily development in the back
- They also had the option of keeping one and developing another.

The Assignment Play: Making Money Off Paper

One of the most **interesting aspects of this deal** was how our **end buyer leveraged an assignment**.

1. **We had the property under contract** at $2,000,000 with a contract price for the end buyer for $3,450,000.

2. **Our buyer then assigned that contract to one of his own entities** at a higher price before closing.

This is a great example of how, when there's **built-in equity and control of the deal**, multiple parties can **creatively structure their profits**.

Key Takeaways from This Deal

- You don't always need to go all the way through entitlement —securing a Preliminary SP was enough to make the property highly valuable to the next buyer.
- Strategic zoning allocations saved twelve to eighteen months for the next developer, making our property more attractive.
- Understanding how assignments and daisy chaining work can create additional profits on deals without ever taking ownership.
- Having control of the contract is key—when you structure the deal correctly, you can maximize your options and create multiple layers of profit potential. This deal demonstrates **the power of strategic entitlements** and how land investors can **add value without getting bogged down in costly site plan approvals**.

CASE STUDY 2: 411 W TRINITY, NASHVILLE TN, 37218 (ADDRESS HAS SINCE BEEN CHANGED)

Purchase Price $1,000,000 | Exit price $4,960,000 (and Some Missed Opportunities)

411 West Trinity was one of the more interesting transactions I've worked on, and it pairs perfectly with the marketing section because it came directly from one of my texting campaigns. It is a combination of a subdivide along with entitlements. This deal took persistence, strategic timing, and quick decision-making to pull off—and while it was an incredible success, there were also a couple of missed opportunities along the way.

The Initial Outreach & Acquisition

Back in 2017, I started texting commercial property owners along major corridors in Nashville. One of the properties I reached out to multiple times over the years was 411 W Trinity. The seller had a "Car Wash Coming Soon" sign up for years, but the project never materialized.

I texted him in 2017—no interest.

I texted him again in 2018—same response.

Finally, in 2019, he replied, "Yeah, I'd sell for $1 million."

I immediately pulled the zoning and did some quick due diligence, and it turned out the site had a mix of commercial and residential zoning across its four parcels. The front two parcels were commercially zoned, and the two in the back were residential.

This was an immediate green light for me because of Nashville's Adaptive Residential Zoning—a zoning that allowed for more density when the majority of a property's frontage was on a major arterial road. By simply subdividing the land, we could increase its value overnight.

As soon as I confirmed this, I sent him a contract immediately at his asking price of $1M, wired earnest money, and sent it to the title company to ensure we had a real deal.

Speed Matters: Closing Before the Market Caught Up

While waiting for the title work, I checked in with the seller a few weeks later. He casually mentioned, **"Man, I'm getting a lot of calls all of a sudden."**

That was a **red flag**.

Texting for land was still a newer strategy at the time, but Nashville's market was booming. If I didn't lock this deal in fast, someone else would swoop in and offer more money.

I immediately called my partners, and within a week, we had each put up $330,000 and some change in cash, got it to the title company, and closed the deal as fast as the title work came back.

Executing the Entitlement Strategy

After closing, we had to decide what to do with the land. Since it was already zoned appropriately, we didn't need a rezoning—but we did need to legally combine the two commercial parcels into one and subdivide the residential ones to make them buildable according to existing zoning.

During the planning process, we discovered that while the commercial zoning technically allowed for short-term rentals (STRs), there was a major restriction—STRs were prohibited within 100 feet of a daycare, church, or school. Since there was a daycare directly next door, this meant we couldn't blanket the entire site with STRs as originally anticipated. Instead of scrapping the plan, we knew we had to be strategic with the site layout to maximize the number of units that could still qualify for STR use. By carefully designing the site plan, we positioned certain units outside of the restricted radius while ensuring the remaining homes would appeal to owner-occupants or long-term tenants. This approach allowed us to create a diverse mix of product types, making the development attractive to both end-user buyers and investors looking for STR-friendly properties.

Our exit strategy included:

- Combining the commercial parcels to ensure maximum frontage for adaptive residential zoning.
- Splitting the residential parcels into four separate single-family lots.
- Keeping the commercial and residential sales separate, since they attract different buyers.
- Listing with a broker at a much higher price than we initially thought we could get.

- Working with our engineer to maximize STR-eligible properties.

Once we had everything platted, we listed the site for sale at around $6 million, knowing the demand for both short-term rental properties and commercial land was skyrocketing in Nashville, and we knew we'd have to work with the end buyer on entitlements, so we didn't go any further with it. Heck, we never even got a survey for it!

We soon received an offer from a buyer we already knew through our network. We finally agreed to a price of $4.8 million, and given the current market, we accepted.

The buyer used the same engineers we had already started working with, making the entire transaction seamless and transparent. It became a team effort, where we helped him maximize the site's potential.

Missed Opportunity #1

At the same time, we knew the property next door was also for sale—off market, but the seller wanted far more than market value. We knew that if this property was assembled to ours, the whole site would be STR eligible. Our buyer knew this and asked if we had talked to them. We told him our story and gave them his contact information. While we could have worked with the daycare owner ourselves and whole-saled it to him, we decided to simply pass the seller's contact to the buyer and let them negotiate directly. We knew if they could make a deal, our deal would have a better chance of executing, and we didn't want to risk jeopardizing a $4.8 million deal just to squeeze out extra profit, which most likely would be minimal since the seller was asking for such a high price.

To me, risk mitigation outweighed the potential gain, so I don't consider this a missed opportunity—just a smart decision to keep our focus on closing the deal we had.

The Final Closing Hurdles & Re-Trade

Over the next year, the buyer worked on grading permits, site plan approval, and resolving access issues. Some key challenges included:

- Retaining wall issues that forced a reduction in unit count.
- Fire access restrictions that further adjusted the site plan.
- A shared easement with Dollar Store next door, which took months to finalize.

Due to these issues, the buyer renegotiated the price down from $4.8 million to $4.4 million, but with a guaranteed close in thirty days.

We took the deal and closed!

Selling the Residential Pads & Missed Opportunity #2

After selling the commercial portion, we still owned the four single-family lots in the back.

At first, we weren't sure what to do with them. We were going back and forth, contemplating whether to build and sell new homes or just sell the lots. However, we had some foresight, knowing that sewer capacity could become an issue later. So, we proactively granted an easement on one of the residential lots so that the commercial developer who just closed on ours could access additional utilities if needed from the street behind us that fronted the residential lots.

This helped remove unknown risks for the new owner while still allowing us to sell the residential lots for top dollar.

Fast forward two years, and we finally decided to list the residential pads for sale without building new homes.

We sold the residential lots for a total of $560,000.

The buyer who purchased them eventually rezoned the four lots into eight—something we could have done ourselves but didn't. At the time, the neighborhood policy didn't allow density, but since it was an area that needed growth, they let him rezone them.

In hindsight, this was a missed opportunity—but it was still a profitable one. At the end of the day, I'd rather make money and learn a lesson than lose money trying to be perfect.

Final Takeaways from 411 W Trinity

- Follow-up is everything. If I hadn't texted this seller consistently for two years, I wouldn't have landed this deal.
- Move fast when the market is hot. We closed as fast as possible to avoid competition swooping in.
- Subdivides add massive value. By combining and subdividing the parcels, we increased the site's marketability and pricing.
- Don't get greedy. We passed on a potential side deal to ensure a smooth close on our $4.4 million sale.
- Think long-term. The easement we granted on the residential pads removed obstacles and made the transaction easier for all parties.

Even with the minor missed opportunity on the residential rezoning, this deal was a home run, turning a $1 million purchase into a $4.96 million exit.

CASE STUDY: THE THREE-YEAR JOURNEY TO CLOSE 839 WEST TRINITY NASHVILLE TN 37218 IN NASHVILLE—AN INFILL TOWNHOME PROJECT

Purchase price $1,500,000 | Exit price $4,150,000.

839 West Trinity in Nashville was one of the trickiest deals I've ever worked on. It took nearly three years to finally come together. This deal had everything: landlocked property, title issues, zoning challenges, wholesaler interference, and lengthy negotiations. But through patience, persistence, and a deep understanding of land entitlement, we made it happen.

How It Started

Back in 2021, when we had shifted our focus to entitling land for development, this particular seven-acre property was on everyone's radar. It was sitting on top of a hill with incredible downtown views in an up-and-coming part of Nashville where everything was changing. Developers and investors were all over this area, and the seller knew it.

The problem? The land was completely landlocked—there was no legal access to a public road. And when you have a situation like that, the land can either be worth a lot of money or absolutely nothing, depending on whether you can secure an easement.

We knew that getting access through the property to the south—which was owned by another developer—was the key to unlocking the value. Without that, it was just a useless piece of land.

The Wholesaler Problem

Like many great properties, this one was first tied up by a wholesaler. Wholesalers often overpromise and underdeliver, locking up deals for prices that don't make sense and then scrambling to find a buyer at a higher price. More often than not, they fail.

This wholesaler had the property under contract for $3 million, which we knew was completely unrealistic. When we spoke to the seller, we asked about the contract terms—how much earnest money was put down and how long the due diligence period was.

This is a crucial part of land investing—understanding whether a contract will actually have the legs to close or if the wholesaler is just running a bluff.

From the details they gave us, we knew for a fact that this deal wouldn't happen. The wholesaler had only six months to close, with very little time for due diligence. Given the encumbrances and access issues, there was no way a serious buyer was going to take that risk.

So, instead of competing for the deal at an absurd price, we waited them out.

Timing is Everything

While the wholesaler's contract ran its course, we kept talking to the city, engineers, and local officials to figure out exactly what it would take to entitle this property. We studied the zoning, environmental factors, slope restrictions, and what approvals were necessary.

At the same time, we built a relationship with the owner of the neighboring property—the one that could grant us access to the land. We knew that as soon as the wholesaler backed out, we needed to be in a position to strike immediately.

Sure enough, one day, we called the seller, and the wholesaler had backed out.

By this point, land values had already appreciated significantly, and what once seemed like a high price now made sense. The seller had been asking for $1.5 million originally, and though we had previously offered around $700,000 to $800,000, the market had changed enough that $1.5 million was now a fair deal.

We locked it up.

The Entitlement Process Begins

With the contract in hand, we began the entitlement process. A few key steps included:

- **Securing Easement Access:** This was critical. We had to wait for the developer to the south to finish their approvals before we could get an easement through their property. This alone took months.
- **Title Work & Quiet Title Action:** This property had some serious title issues, including unclear ownership due to past

owners passing away without clear heirs. We had to file a
quiet title action, which meant sending out notices and holding
court hearings. If no one contested the ownership within a
certain period, we could proceed with the sale.

- **Working with an End Buyer:** Once we had everything in
  motion, we secured a contract with a large builder who was
  interested in developing the site. Their contract was contingent
  on site plan approval, meaning they needed the city's final
  green light before they could move forward.

- **Financing & Guarantors:** Our buyer also had to raise capital
  for the down payment and bring in a guarantor—someone
  with the financial backing that a bank would lend to. This
  process took time because large-scale developments often
  require multiple parties to sign off.

The Final Hurdle: A $1.5 Million Problem

Just three weeks before closing, we hit a major snag.

The title company had made an error in the original closing documents,
meaning our title wasn't clear. That meant our buyer couldn't close
unless we went back to the original seller and got them to sign a
corrective deed.

If you've ever dealt with land sellers before, you know that once a
seller sees you're rezoning or reselling the property for a profit, they
suddenly don't want to cooperate.

This seller was no different.

We had already paid them $1.5 million, and now we had to go back to
them asking for one more signature. And, of course, they wanted more
money to sign off on it.

So, we had two options:

1. **Fight them legally, which could take months and delay everything.**
2. **Pay them off and get it done.**

We chose option two, wrote them a check, got the correction signed, and closed the deal.

Lessons Learned

This deal took three years from the first conversation to closing. It was frustrating at times, but it was also one of the most rewarding projects we worked on.

- **Patience is key.** Sometimes, you just have to wait out bad contracts and let the market work in your favor.
- **Relationships matter.** If we hadn't kept in touch with the seller and the neighboring landowner, we wouldn't have been in a position to take this deal when the wholesaler backed out.
- **Title issues can make or break a deal.** If you don't have clean title, you don't have a deal. Period.
- **Easements are everything in land deals.** If you don't have access to a public road, your land is worthless.
- **Know your exit strategy.** We locked up the land knowing we could entitle it and sell it for a premium—and that's exactly what we did.

This deal was a perfect example of why land entitlement is so valuable. We took a property that was nearly worthless due to access issues and turned it into a multi-million-dollar development site.

And that's why I do what I do.

EPILOGUE: THE ROAD TO MASTERY

L and development is not just about transactions, contracts, and approvals—it's about vision, persistence, and strategy. By now, you should have a deep understanding of the entire process, from finding and evaluating land to structuring deals, entitling property, and ultimately exiting for a profit. But knowing is only half the battle—taking action is what separates those who succeed from those who remain on the sidelines.

One of the biggest takeaways from this book is that there is no single formula for success. Every deal is different, every city has its own regulations, and every seller or buyer has their own motivations. The ability to **adapt, problem-solve,** and **build relationships** will take you further than just understanding zoning codes or market trends.

You've learned how to **identify undervalued land,** leverage GIS mapping, and build a strong marketing plan to source off-market deals. You now understand how to **negotiate contracts, structure financing,** and **navigate the entitlement process.** You've been introduced to the **importance of due diligence, working with the right team,** and **understanding city politics.** And most importantly, you now have the tools to confidently analyze, structure, and execute deals in a way that minimizes risk and maximizes reward.

WHAT SETS THE BEST DEVELOPERS APART?

The best developers don't just look for land—they **create value where others don't see it**. They think long-term, remain flexible in their approach, and understand that time is their biggest asset. It takes patience to work through city approvals, rezoning, or land entitlement. It takes persistence to work through challenges when a deal seems like it's falling apart. And it takes confidence to invest capital and time into a project that may not pay off for years.

But that's why **so few people do this successfully**. The uncertainty, risk, and long timelines scare most investors away. They want fast flips, easy deals, and predictable returns. Yet the developers who **master this business stand to gain the most**, building generational wealth by unlocking land's hidden value.

WHAT'S NEXT FOR YOU?

Now that you have this knowledge, the next step is **putting it into action**. Start by identifying a market, researching land opportunities, and networking with the right people—**agents**, **engineers**, **city planners**, and **experienced developers**. Get comfortable reading zoning codes, using GIS tools, and analyzing potential exits. And when you find the right deal, apply everything you've learned here to structure a winning transaction.

FINAL THOUGHTS

I've shared real-world case studies, strategies, and insights that took me years to learn—often through trial and error. My goal is that you walk away from this book **not just inspired, but prepared**. The land development business isn't easy, but it **is predictable if you follow the right steps**. Stay focused, keep learning, and **surround yourself with the right people**.

Most importantly—**take action**! The best deals don't wait, and the biggest opportunities go to those willing to **put in the work, take calculated risks**, and **execute consistently**.

This is just the beginning. See you in the field!

REFERENCES

James, Geoffrey. August 29, 2016. "Motivational Quotes From 49 Intensely Competitive People." Inc.com. Accessed March 21, 2025. https://www.inc.com/geoffrey-james/ motivational-quotes-from-49-intensely-competitive-people.html.

ABOUT THE AUTHOR

Greg Farricielli is a seasoned real estate investor, developer, and financial strategist with a deep-rooted expertise in land acquisition, entitlement, and development. His career began in 2015 as a real estate agent in Nashville, where he quickly built a reputation for identifying high-value investment opportunities and structuring deals that generated substantial returns.

Greg started by acquiring rental properties and leveraging strategic market insights to build a cash-flowing portfolio. As his expertise grew, he expanded into real estate syndications, successfully raising capital for multifamily investments. Recognizing the power of creative financing, he ventured into hard money lending, providing funding for investors and developers while sharpening his understanding of risk, leverage, and deal structuring.

His ability to analyze market trends and identify undervalued assets led him to specialize in land development—transforming raw or underutilized land into high-yield residential and mixed-use projects. Through his partnership, **Rhythm Development**, Greg has played a pivotal role in shaping Nashville's growth, as well as developments in surrounding areas. His projects focus on maximizing land value through rezoning, infrastructure planning, and strategic partnerships with national home builders.

Now based in **Tampa Bay, Florida**, Greg applies his expertise to new markets, targeting distressed and high-potential properties in **Pinellas** and **Hillsborough Counties**. With a focus on **lot-value properties**,

storm-damaged assets, and **redevelopment opportunities**, he leverages his experience to structure profitable deals that create win-win outcomes for investors, developers, and local communities.

Beyond real estate, Greg is an advocate for **financial freedom, faith-driven entrepreneurship**, and **community impact**. His mission is to develop properties that not only generate wealth but also contribute to thriving, sustainable communities.

In this book, Greg shares the **hard-earned lessons**, **strategies**, and **financial frameworks** that took him from an agent selling homes to a land developer creating multimillion-dollar opportunities. Whether you're a real estate investor, builder, or entrepreneur looking to break into land development, this book will provide the insights and roadmap to navigate the complexities of the industry—without making the costly mistakes that most newcomers face.

www.ingramcontent.com/pod-product-compliance
Lightning Source LLC
La Vergne TN
LVHW051413080426
835508LV00022B/3069